A Sailor's Song: Lost Love Letters of World War II
By Larry Schnell

Printed in the United States of America.
Published by Intrada Press LLC

© Copyright 2025 by Larry Schnell

For bulk purchases or to carry this book in your retail establishment, library, university, website, or school, please contact the publisher at LarrySchnellfl@yahoo.com.

ISBN: 979-8-218-73026-0

Intrada Press LLC
PO Box 51
Van Hornesville, NY 13475 USA

A Sailor's Song:

Lost Love Letters of World War II

by Larry Schnell

Reviews and Radio Interviews

A Sailor's Song: Lost Love Letters of World War II is a welcome look at
the 40s, America in the world war, one family's separation and survival
and the Coast Guard. It is a nostalgic and poignant look that makes me
proud to be an American. It was a time, when our people, put aside their
political, social and economic issues and fought together to lead the world
to triumph over fascism. Based on his father's correspondence with his
mother while he served in the Coast Guard, Schnell's readers are fortunate
to have this story of life at sea and the home front during the war. *A
Sailor's Song* is solid cultural history and, more important, a very readable
story that anyone interested in the Greatest Generation will enjoy.

Dr. Robert L. Gold, PhD
Historian, Professor, Author

This is a lovely gem of a story about band music and love letters
exchanged in wartime between the author's parents Arthur and Florence
Schnell. Art, a music teacher and trombonist, enlisted to serve in the
Manhattan Beach Coast Guard Band during WWII. The author blends
the love story between his parents, separated by Art's wartime deployment,
with the rigors of working on a naval transport ship that, from 1945-
1946, sailed through the Pacific, Indian and Atlantic Oceans, and
circumnavigated the world to transport troops, supplies and mail. Art's
carefully coded letters secretly reveal his geographic locations to his wife
to evade the watchful eyes of the censors. A blend of narrative nonfiction
and historical memoir, *A Sailor's Song: Lost Love Letters of World War II*
provides the reader with detailed information of the role of musicians in
the U.S. Coast Guard, and is a glimpse into the love story of two people
separated by war yet determined to keep their romance alive.

Susan Waller Lehmann
Author, Private Investigator, Journalist

Jacksonville Public Radio
First Coast Connect WJCT 89.9's recent review of *A Sailor's song: Lost Love
Letters of World War II* focuses on the book's relevance to sailors during
World War II while highlighting the role of music in the war. The author's

parents, both musicians during the war, shared detailed information about the music beloved of Coast Guard/Navy musicians as well as the American public. "What really spoke to them was this really sometimes overblown but wonderfully romantic music of that era and of the sound of big bands, which is what your father and his colleagues enjoyed performing because that is what the soldiers and sailors liked to listen to," she noted during the live interview. The interview is available at https://www.youtube.com/watch?v=upDOUjmBDng.

Albany Times-Union
Local books worth gifting for young, older readers alike, with Jack Rightmyer, Times Union. "This touching story began with the author finding a box of forgotten war letters from his Coast Guard sailor father and his wife, hidden in his parents' attic.... This story captures the 1940s perfectly, and also includes historical records, news articles from the time and even deck logs. These letters kept their love alive, and it also provided some hope that he would survive the war, and they would have a long and loving life together."

Ithaca College Memorial Day Post
A Sailor's Song: Lost Love Letters of World War II was featured in an Ithaca College Instagram Memorial Day 2025 post.
"When Larry Schnell found his IC alumni parents' WWII letters—between Arthur Schnell '40, a Coast Guard musician, and Florence Schnell '39, a music teacher and performer—their words revealed a hidden code, a love story, and the power of music in wartime. He turned them into a book: A Sailor's Song: Lost Love Letters of World War II. Dedicated 'For the musicians of war, inspiring troops and home-front families, healing the injured, honoring the dead, singing yet unsung.'"

Illinois Public Radio
A Sailor's Song: Lost Love Letters of World War II was featured on IPM News, Champaign, IL, April 4, 2025, with Anna Koh on "217 Today: How a hidden box of love letters became a historical memoir of WWII." "In today's deep dive, we'll learn about a new memoir from local author Larry Schnell that explores the little-known role of music in war."

Savannah Radio

A Sailor's Song: Lost Love Letters of World War II was featured on WRUU Savannah, April 19, 2025, in an interview with P.T. Bridgeport for his program "When the Moon Sings." Mr. Bridgeport, whose father was a prisoner of war, provided an excellent forum for discussion of the book, focusing on military and home-front morale. He notes, "You are providing people with a sense of what it was like to be in the war both from the military perspective and the civilian perspective."

Community Radio Ithaca

WRFI Community Radio News aired a two-hour documentary on *A Sailor's Song: Lost Love Letters of World War II*. How Deep Is the Ocean: Larry Schnell on A Sailor's Song by Roger Kimmel Smith aired August 8, 2025. The two-hour broadcast features Smith's extended interview along with 1940s music, news clips, and other historic audio. Arthur Schnell was a Coast Guard musician who described his wartime experiences in letters to his wife, Flo Schnell, a civilian. The roles played by musicians during wartime is the theme of this documentary.

News-Gazette

News-Gazette in Champaign-Urbana review in 'A Sailor's Song,' Coast Guard brassplayer made waves across country, with Rob Le Cates, The News-Gazette, Champaign-Urbana. "Local author spotlights music, war and power of letters in new book."

The Daily Star

With Ella Connors, The Daily Star, Oneonta, November 7, 2025. Local author spotlights music, war and power of letters in new book. In March 1945, Schnell said, his father and about a dozen other musicians were shipped to California to serve on the U.S.S. *General A.W. Greely*. They learned their commander George W. Stedman Jr., a former cruise ship captain, loved music. "He believed that music would be good for the young troops who were heading to the Pacific because they were planning to invade Japan," Schnell said.

Table of Contents

Dedication..7

Acknowledgments...11

Preface...13

Chapter 1 Instruments of War....................................19

Chapter 2 Message in a Box.......................................25

Chapter 3 Music of War..31

Chapter 4 Whispers of War...41

Chapter 5 West to the Pacific.....................................51

Chapter 6 Home-Front Hardships..............................65

Chapter 7 San Francisco..67

Chapter 8 Gateway to the Pacific War........................71

Chapter 9 Commander George W. Stedman Jr. 83

Chapter 10 Loading the Ship 89

Chapter 11 Shakedown..95

Chapter 12 Somewhere in the Pacific.......................109

Chapter 13 Marriage and War....................................121

Chapter 14 Australia..131

Chapter 15 Censorship..141

Chapter 16 War Letters...153

Chapter 17 Coming Home...165

Chapter 18 France and the Pacific............................171

Chapter 19 Bringing Them Home.............................181

Chapter 20 After the War..189

Notes..197

Bibliography..205

Appendix...212

About the Author..223

For the musicians of war, inspiring troops
and home-front families, healing the injured,
honoring the dead, singing yet unsung.

Acknowledgments

I wish to thank those who volunteered to help me research, search archives, and complete A Sailor's Song. The extensive variety of topics required me to tap the resources of many people with varied expertise, whose help I gratefully acknowledge. I am grateful to Coast Guard historian William H. Thiesen who directed me to the many resources and locations of records on the subject of the Coast Guard's Manhattan Beach Training Station. Much thanks to historian and author Dr. Robert Gold, who brought history to life with his books; his inspiration and guidance are gratefully acknowledged. Kimberly Guise, who studied thousands of war letters archived in The National WWII Museum, documented their value as primary sources for understanding World War II and helped me realize the historical value of the letters my parents left me. Herkimer College librarians Ann Prior and Stephanie Conley sifted through historical articles and records online to find essential information buried in the archives of history. Christopher Sturcke with the Jordanville Public Library found obscure books with valuable information on censorship in World War II. Brother Bill Schnell lent me a collection of informational booklets provided to troops and crew on the U.S.S. *Greely* and preserved one of the "code letters" that defines the encryption my parents used in correspondence to identify location at sea. Much thanks to my cousin David Cape and wife, Eileen, for hosting my stay in Washington, D.C., while I reviewed records in National Archives 1 and 2. Kat LaMons, director of Marcinson Press, provided valuable guidance and editing for the original publication. I am thankful that my parents wrote detailed letters about music and love that provide an enduring perspective on war and music, and that they preserved them along with a trove of war memorabilia. Special thanks to my partner Debra Donahue, who listened to, read, and edited drafts of *A Sailor's Song*, often under duress.

Preface

President Roosevelt enacted the Office of War Information in 1942 to create and depict an image of World War II, portraying home-front struggles and battlefield heroics. Images in magazines, movies, newsreels, and on posters defined the war for the American people. The OWI tapped America's artistic talents to create powerful and patriotic images, stories, and films. OWI Director Elmer Davis realized that the most effective producers of propaganda were in Hollywood. "The easiest way to inject a propaganda idea into people's minds is to let it go through the medium of an entertainment picture when they do not realize they are being propagandized," Davis wrote.[1]

Roosevelt agreed and authorized the formation of the OWI's Bureau of Motion Pictures, which would define and regulate the content of Hollywood's output during the war. By the end of 1942, nearly all major studios submitted scripts and storylines to the bureau for approval, adhering to the standard, "Will This Picture Help Win the War?" The OWI's Manual for the Motion Picture Industry became the standard most Hollywood producers adhered to in writing scripts and filming subjects.

Movie makers were the most prominent contributors to the OWI's propaganda program, but they were not alone. Artists, celebrities, and photographers acquiesced to the OWI's authority and helped the federal government cast the war in a favorable light. While the OWI crafted the release of information, the Office of Censorship under veteran newsman Byron Price restricted the release of information and prevented potentially damaging information from getting into enemy hands. While regulating stories from news media, Price enacted a policy of voluntary compliance. Reporters and editors submitted their reports to censors and abided by their decisions. With few exceptions, self-regulation of news media was a success. With the Office of War Information crafting favorable stories and images, and the Office of Censorship withholding potentially damaging news and information, the federal government oversaw a propaganda program un-precedented in American history.

Music was the lone creative genre the federal propaganda program failed to control. The OWI's Music War Committee was charged with producing and popularizing appropriate war music and encouraging musicians to play patriotic songs. Top musicians had emerged from the wildly popular big-band era; band directors such as Glenn Miller, Tommy Dorsey, Artie Shaw, and vocalists Rudy Vallée, Judy Garland, Doris Day, Marlene Dietrich, Dinah Shore, and many others entered the war years with a trove of ro-mantic and sentimental songs for an audience with an insatiable appetite for slush. Rarely did a patriotic song rise near the top of the charts, and those that did fell quickly. Military band leaders like Glenn Miller with the Army, Artie Shaw with the Navy, and Clare Grundman with the Coast Guard knew what kind of music the troops wanted to hear. Even with a military mandate for patriotic songs, their repertoires were laced with sentimentality that had nothing to do with war. The Army's *Hit Kit*, a pocket-sized book of patriotic songs the Army felt would inspire troops in battle, failed in a year. The Music War Committee failed almost from its inception. Despite the best efforts of the military and OWI, the federal government failed to change the musical tastes of the American public, top musicians, or the soldiers and sailors they set out to inspire.

Musicians contributed to the war effort in their own way, playing pop-ular songs for families at home, for troops heading to the front, and for soldiers recovering from battle. While the military was not fully prepared to address the psychological damages of war, psychology counselors were beginning to understand the value of music as an anodyne. The positive effects of music in World War II laid the groundwork for the de-velopment of modern music therapy, the practice of using music to treat people with mental conditions and trauma. Concordia College's memory project explored the value of music in helping troops before and after battle. "Music played an integral role in the lives of the soldiers. Not only were they kept distracted from the aspects of their daily lives in combat, but they were also greatly assisted by music upon their return from the war. With the development of the music therapy field at this time, one can see that the troops returning had a better, more positive outlook on life than in previous wars and music was a large contributor."[2]

The story of World War II is too vast to be told in its entirety. The story of music in World War II has filled volumes. This is the story of the Coast Guard's Manhattan Beach Training Station bands as told in official documents and letters between Arthur Schnell, a Coast Guard musician, and his wife, Florence, a music teacher and performer. Throughout much of the war, Art and the Manhattan Beach Coast Guard musicians played in multiple bands for bond rallies, Mutual Radio broadcasts, officers' balls, ship commissionings, and other events related to war. During the last year of the war, most of the Manhattan Beach musicians were assigned to the U.S.S. *General A.W. Greely*, a new troop transport ship that would carry troops between the homeland and the Pacific war. Initially, the Manhattan Beach musicians were concerned that their talents would be sidelined, yielding to other war endeavors. But George W. Stedman Jr., commander of the *Greely*, had no such plans. Formerly a cruise ship captain, Stedman recognized the value of musical entertainment. He transformed the *Greely* from a warship to a musical ship for the comfort of troops before and after battle.

The story of music on the *Greely*, told through more than 150 letters my parents exchanged, records in the National Archives, and ship publications, provides a glimpse of the war from perspectives aboard the ship and at home. Portraits of a dozen musicians and their wives tell of the struggles many thousands of war couples endured during separation. The letters document the rise of a novice musician in non-musical boot camp to a prominent musician in Coast Guard bands that entertained and inspired on base, in the streets, over the radio. They portray the bands' transition from the security of a land-based assignment to sea service on the *Greely*.

The National WWII Museum in New Orleans is a trove of personal perspectives on the war, preserved in letters from servicemen in distant fronts and families at home. The letters, like those exchanged between Arthur and Florence Schnell, are primary and enduring records of war from survivors and from those who lost their lives even as their last letters traversed the globe in the holds of ships.

Welcome aboard the U.S.S. *General A.W. Greely*. Find a seat on deck near Hatch 5. The concert is about to start.

A Sailor's Song: Lost Love Letters of World War II portrays some of the Coast Guard's many wartime contributions that often go unnoticed in the vast archives of World War II. In more than two centuries of service, the Coast Guard has performed naval combat operations in two world wars and subsequent conflicts while embracing civilian responsibilities of maritime safety, port security, licensing, and inspecting of vessels. The Coast Guard motto Semper Paratus (Always Ready) is not just a slogan. It reflects the Coast Guard's preparedness to engage in naval warfare again should it be required for national security. These operations would be carried out even as the Coast Guard maintains responsibilities associated with peacetime, as it did during World War II.

The Coast Guard's role in national security and maritime safety has its origins in the early days of the republic. The Revenue Marine and the Revenue Cutter Service, as it was originally named, is the longest serving branch of the military. Under authority of the Treasury Department, the service was the only armed forces afloat, predating the Navy by eight years. Through the nineteenth and early twentieth centuries, the roles and responsibilities of the Revenue Cutter Service expanded to meet the needs of a growing nation. The Coast Guard received its present name in 1915 when Congress merged the Revenue Cutter Service with the U.S. Life-Saving Service. In its new role, the Coast Guard assumed responsibility for dozens of coastal stations, absorbing more than 5,000 personnel, as well as ships, depots, and district offices as it took responsibility for maintaining 30,000 aids to navigation.

In World War I, the Coast Guard assumed military responsibilities, providing nearly 9,000 Coast Guard men and women who fought under command of the Navy as warship commanders, troop-ship captains, training-camp commandants, and naval air-station commanders. Recognition of service included two Distinguished Service Medals, eight Gold Life-Saving Medals, fifty Navy Cross Medals, and many foreign honors.

As the next war in Europe threatened to enmesh the United States, Coast Guard cutters, destroyer escorts, and frigates patrolled the North Atlantic, where Nazi submarines prowled the waters to attack allied shipping. By the spring of 1941, the Coast Guard had seized scores of Axis merchant ships, routed German agents from Greenland, and rescued several thousand mariners who survived Nazi torpedo attacks.

As the United States faced the threat of war, President Roosevelt on November 1, 1941, transferred authority of the Coast Guard from the Department of Revenue to the Navy, where it prepared to perform naval combat operations in addition to traditional services. Naval engagement quickly followed. The cutter *Icarus* sank a U-boat, capturing German mariners and delivering the first enemy prisoners to the U.S. continent. *Icarus* captain Lt. Maurice Jester was awarded the Navy Cross. Over the course of the war, Coast Guard crews sank eleven U-boats.

The Coast Guard was fully engaged when the Japanese attacked Pearl Harbor. While cutters fired antiaircraft barrages against Japanese aircraft and patrolled the harbor for submarines, the Coast Guard's first combat casualties were recorded on opposite sides of the globe. Off Iceland, an attack on the cutter *Alexander Hamilton* resulted in twenty-six dead and fifty-six wounded. That same day, the Coast Guard troop transport U.S.S. *Wakefield* was refueling in Singapore when a Japanese bomb penetrated the deck, exploding in sick bay, killing four Coast Guardsmen. The global war also threatened the U.S. coast, where lifesaving stations and lighthouses became defense structures. Nearly 2,500 Coast Guard personnel patrolled thousands of miles of coast on foot, on horseback, in vehicles, guarding U.S. ports and foiling plots to land German spies.

In the Pacific war, the Coast Guard was engaged in the battle of Guadalcanal, transporting Marines, supplies, and equipment to the island. In a dangerous rescue operation evacuating Marines ambushed by Japanese, Signalman First Class Douglas Munro directed a flotilla of landing craft to a successful rescue. Munro died by gunfire in the process and was posthumously awarded the Medal of Honor. The Coast Guard participated in similar operations throughout the Pacific as well as North Africa and Italy while continuing rescue operations of torpedoed ships in the Atlantic. By the war's end, Coast Guard cutters and aircraft rescued nearly 1,000 Allied and enemy survivors in the North Atlantic, another 1,600 along the American coast, and 200 in the Mediterranean.

Coast Guard officers participated in planning Operation Neptune, the Navy's role in Operation Overlord. Coast Guard crew joined the assault on Normandy, commanding assault transports, cargo ships, and landing crafts along Omaha and Utah beaches. Cutters rescued 1,468 troops from coastal waters and offshore victims of sinking ships.

With the collapse of Germany, May 8, 1945, Hitler's successor, Grand Admiral Karl Dönitz, broadcast the order to surrender. Cutter *Argo*'s skipper, Lt. Eliot Winslow, led the unit of six ships that accepted the surrender of German submarines and warships. During the war, the Coast Guard had transported hundreds of thousands of troops to the Pacific theater, some aboard the U.S.S. *General A.W. Greely*, as portrayed in *A Sailor's Song: Lost Love Letters of World War II*. Following the surrender of Japan, the Coast Guard began repatriating troops in Operation Magic Carpet. The U.S.S. *Greely* joined the repatriation effort, returning some of the longest-serving troops in the Pacific theater. In the ensuing peace, the Coast Guard was returned to the Department of Revenue and continued its peacetime operations, returning to naval combat in conflicts in Korea and Vietnam.

When terrorists struck the World Trade Center on September 11, 2001, Coast Guard personnel immediately assumed the role of saving lives and providing security. With bridges and tunnels in Manhattan closed, the Coast Guard mobilized hundreds of civilian vessels to evacuate 500,000 people in eight hours, the largest maritime evacuation in history. In support of Operation Noble Eagle, thousands of Coast Guard men and women in the reserves and auxiliary were mobilized to support the largest homeland defense and port security operations since World War II. The 2001 terrorist attacks reshaped the Coast Guard in its new role under Homeland Security, where it maintains its fundamental principle, "Semper Paratus."[3]

CHAPTER ONE

Instruments of War

A dozen Coast Guard sailors arrived at their new station in Alameda, California, prepared to take up weapons of choice—musical instruments—against the Japanese. The Manhattan Beach Training Station musicians would soon board a ship destined for ports in the Pacific. Their new role in the war began with chaos and uncertainty on arrival at the Naval Air Base in Alameda in early March 1945.

There was little to do at first other than muster twice a day, play cards, wait, wonder, worry. The band members hoped they would, in some capacity, function as a band, but they had no band leader. Surely, they were not assigned to the portal of the Pacific War to just play music. They passed the time and tried to thwart uncertainty with dinners in Oakland, second-rate movies, sightseeing, and card games. Several musicians spent a comfortable night as guests of the San Francisco band, one of ten regional Coast Guard bands formed during the war. Their gratitude to the San Francisco band was tempered by a nagging question: Why was the San Francisco band continuing its musical war role while the Manhattan Beach band was gutted and its leading musicians, shipped out? Was it the image of the white dress uniforms and polished shoes they wore as they marched and played in Manhattan streets while bloodied soldiers fought in grim conditions? Was it a sense of privilege they assumed and at times flaunted as they horsed around far from combat zones? Was it triggered by an inappropriate incident involving a drumstick that occurred while the band was being reviewed by an admiral a week or so before they were ordered to California? While band members speculated and differed over the causes of their transfer, they ignored the Coast Guard's preeminent reason—the war.

They left wives behind in New York with barely a goodbye. The wives quickly bonded as a support group to endure the absence and anxiety over their husbands' new role in the war. The musician-sailors bonded over the confusion and anxiety of their new assignment.

The Alameda Naval Air Station was created just in time for World War II and expanded throughout the war to accommodate air units, carrier groups, supplies, Naval personnel, and this new crew of Coast Guard musicians. The Manhattan Beach band musicians settled into the barracks while they awaited word of their new roles as the war dragged on. They learned they would serve on the U.S.S. *General A.W. Greely*, a new ship of which they knew little and even less about its destinations. What would band members do on the *Greely*, play music, fire guns, man landing vessels, attack Japanese ships? Rumors flourished amid news of the Coast Guard's and Navy's roles in the Battle of Iwo Jima, in progress as the musicians arrived in San Francisco. Some 7,000 Marines died taking the island. Was that a foreshadowing of their mission? Were they to play a role in taking Okinawa, the next steppingstone en route to the Japanese mainland? If the *Greely* were to serve as an attack transport—a rumor prompted by the 5-inch guns fore and aft—she could be a high-profile target for artillery on Japanese shores. As reliable information seeped in, band members learned the *Greely* was not an attack transport. She was a troop transport, a ship that delivered troops to staging areas, not beachheads.

The musicians' relief was brief. Where were the staging areas? Rumors flew furiously. Some said the destination was India, which involved sailing through hostile waters of the Pacific and Indian oceans. Others said Australia, a relatively safe location if the ship could evade torpedoes along the way. Musician Art Schnell, an optimist at this point, believed they were headed for Hawaii. Or maybe, if Germany surrendered, the *Greely* would sail the Atlantic, bringing troops home from Europe and docking in New York Harbor not far from wives of the *Greely* musicians. In a flurry of letters between San Francisco and New York, wives and musician-sailors exchanged hopes and fears while they sought reliable information amid the uncertainty of war.

Finally, a band leader showed up and called a rehearsal, relieving musicians of idle time to churn the unknown into anxiety. Harold Brody had to get the band in shape quickly as the musicians would perform for the *Greely* commissioning, a formal event attended by officers of the Navy, Army,

Coast Guard, and Marines as well as political dignitaries. The band sounded ragged during the first rehearsal, but the Manhattan Beach musicians were among the best and would get it together for the big event. After that, they would play for crew loading the ship. As they helped prepare the *Greely* for its maiden voyage, they observed and shared bits of good news—musical instruments were among the cargo.

Band members did not know how they came to be known as the *Greely Grenadiers*, but after it was mentioned at a performance and published in the ship's newsletter, it was official. Although some musicians did not like the name, they liked the implication that the band would play while the ship was at sea, that music had a role in war. They would soon learn that soldiers awaiting deployment could enjoy calming or inspirational music. That same music could soothe the minds of soldiers returning from years of battle.

By the end of the *Greely*'s voyages, after destination rumors had been replaced by travelogue, after fresh troops had been delivered and war-weary troops returned to their families, the *Greely* would earn a reputation for distinguished voyages and service to veterans.

By the war's end, the *Greely* had completed a circumnavigation via India, three additional trips to India and other Pacific destinations, and two transatlantic crossings. A *Greely* newsletter[1] proclaimed the ship's first trip to Pacific destinations the longest maiden voyage for a warship—almost 13,000 miles—and the longest voyage for a troop transport. Many of more than 10,000 passengers on return voyages were among the longest serving troops in the war. Returning from France, she repatriated the 2nd Infantry 3rd Division (Red Diamond) after its troops helped drive the Nazi armies out of France and back to Germany. Returning from her second Pacific voyage, the *Greely* brought home Merrill's Marauders, who fought in the jungles of Burma, the Flying Tigers, after almost four years in China, the U.S. Army engineers who built the Ledo Road through Burma, the Kachin Rangers who fought the Japanese with help from indigenous people, and other service groups. These famed fighters debarked the *Greely* in New York to the cheers of thousands and to the music of the *Greely Grenadiers*, the band that earned her recognition as the only military ship to have its own band aboard from the day of commissioning.

The *Greely* also delivered mail—ship to shore and shore to ship, thousands of pounds on each voyage. Letters were the only means for sailors and soldiers to keep in touch with loved ones. The Manhattan Beach musicians joined millions of Americans writing letters that were packed in mail sacks and transported throughout the world in the holds of ships like the *Greely*. Couples struggling to maintain a marriage in the uncertainty of war relied on letters to share their hopes for the future, for themselves, and for a nation. Letters from troops in war zones encouraged optimism for loved ones as they portrayed a lifestyle as remote from the reality they left behind as the distance separating them. Those in combat found comfort, encouragement, and hope in letters shipped to distant military bases and battlefields with the hope that the receiver would be alive to read them.

Kimberly Guise, senior curator at The National WWII Museum in New Orleans, studied thousands of letters to and from troops separated from families by World War II. Letters are primary sources that tell enduring stories of the personal side of the war. "These singular resources will survive far beyond the time when the last WWII veteran has passed," she wrote. "The information they provide is immediate, requiring no instruction about how to interpret them. They speak for themselves. Each piece tells us something new, adding a unique personal vision to the immense story of World War II."[2]

The exchange of letters between Arthur R. Schnell, Musician 3rd Class, and wife, Florence, reveals much more than the personal aspects of war. The couple was committed to a life of music, and the information they exchanged reflects the importance of music in the military and on the home front. Flo's letters describe a woman taking charge of a range of issues from disentangling her 1933 Plymouth coupe from a fender-bender to managing a musical career in the absence of her husband. Art's letters provide an intriguing inside look at the role of music in war, on a military base, and on a ship. His letters describe the personalities, talents, and temperaments of the musicians who turned passion into patriotism. Together with historical records, the *Greely* War Diaries and Deck Logs, newsletters, and news articles, these letters portray the importance of music in supporting the war, troops, and home-front families.

While millions of war letters between loved ones were written, mailed, and shipped, Arthur Schnell received the first letter from his wife on March

14, 1945, more than a week after arriving in San Francisco. "This afternoon I received the first letter that anyone in the band has received since coming out here, and was I proud, which just proves that my darling is the best 'Mousie' (the couple's term of endearment)," he replied.

In early April, musicians were ordered to pack their bags, as they were leaving in twenty minutes for Treasure Island to train and prepare for an accelerated shakedown cruise. Letters would be censored, Art warned. "Naturally while we are on board, our mail will be censored so remember the code," he wrote in a letter mailed from a Post Office in San Francisco beyond the eyes of censors. When I first read the letters, I had no idea what code my father was referring to.

During World War I, the Coast Guard suffered the greatest loss of life due to naval combat when the cutter *Tampa* was torpedoed, killing all 131 persons on board.

CHAPTER TWO

⚓

Message in a Box

My mother died in December 2007 in the home she and my father built in 1952, their four-plus decades in the small community of Van Hornesville, a reflection of their love of Upstate New York and of music. Both had taught music at Owen D. Young Central School until retirement in 1974, when they became snowbirds, living between Van Hornesville and Vero Beach, Florida, playing music, canoeing, and hiking in the rich natural resources of Upstate and what was left of Florida's natural environment. After my father died in 1996, my mother continued to live in both houses until her death. My sister, Penny, arranged a memorial service at the house in Vero Beach a few hours from my home at the time. A few days after the service, we began looking through the house for items of significance. I explored the attic and was surprised to find it almost empty, as my parents tended to save things. I focused on the only visible item, an old cardboard box, worn and dusty, sealed with tape that long ago had lost its adhesive.

Downstairs, I opened the box and was both surprised and thrilled to find it full of letters, most from one parent to the other, dating as early as 1938, when they studied music together at Ithaca College. Many were addressed to or bore the return address "A.R. Schnell Mus. 3/c" (musician third class) from the war years. My father's letters, each in its envelope, were arranged chronically, tied with a string. My mother's letters were not in envelopes but were bound in chronological order by rusty staples.

Precious history was surely contained in these letters, history of a loving couple separated by war, history of music during war. I had a personal interest in my parents' early years of marriage and of the war years, but there was much more in the letters, as I was to learn years later. Carefully I opened a few envelopes, glanced through some letters, and returned them to their envelopes when I realized the vast amount of information they contained.

Even at glance, I realized the communication between two consummate musicians documented the importance of music in war, in bond sales, morale building, national identity, and patriotism. My father served in the Coast Guard, performing at the Manhattan Beach Training Station. Toward the end of the war, some musicians of the Manhattan Beach Coast Guard Band were shipped out to California to serve on a troop transport ship, the origin of many of my father's letters. The letters provided a detailed portrait of the Manhattan Beach Coast Guard Band, one of the regional Coast Guard bands formed for the specific purpose of supporting the war.

I did not have the time to carefully review, catalogue, and understand them. And I was not yet prepared to wade through the headwaters of my parents' marriage. As the box was falling apart, I considered discarding it and storing the contents in a new box when I recalled a book by Marshall McLuhan, *Understanding Media: The Extensions of Man*, in which he introduced the phrase and concept "the medium is the message."[1] One element of his argument was that media research in the 1960s focused on the content of television shows while ignoring the simple fact that a box—the medium—emits sounds and images in most living rooms in the country, the focal points of families. That scenario is the primary message. I studied the box in my hands. Surely there was a message in that medium and in the paper and handwriting inside, regardless of the content of the letters. Under the attic dust, the box seemed coated with wax. An inscription read,

DURR'S SLICED BACON
1 LB LAYERS – 12 LBS.NET
CITY OF UTICA INSPECTED AND PASSED.
C.A. DURR PACKING CO., INC. UTICA, NY

Inspected and passed? Some of the envelopes that contained letters my father wrote from the ship were stamped "PASSED NAVAL CENSOR." I wondered how my parents felt about his intimate expressions being reviewed by a military bureaucrat. We're not talking about bacon. It would be six years before I would find out and learn the code my father referred to in one of his letters.

After the war, my parents settled in Van Hornesville, an hour's drive from Utica, New York, where they built a house in 1952. At some time in Upstate New York, the bacon box became the repository. The wax coating may have helped preserve the writing paper, which itself told a story, although an insect got into the box and ate part of a V-mail letter—a military form of correspondence that allowed censors to easily review the content. It is unclear why and when the box was relocated to Vero Beach. Were the contents so precious that my parents carried the box back and forth? Why not leave the letters in New York with the other boxes of memorabilia? A huge scrapbook almost two feet rectangular contained records of my father's time on the ship, the U.S.S. *General A.W. Greely*. His Coast Guard dress and work uniforms also were in the attic along with leggings and a ship hammock, all marked in indelible ink "Arthur R. Schnell." Newspaper articles stored in folders blared, "ROOSEVELT ON FINAL JOURNEY," "ALLIED SHOCK TROOPS CONTINUE SWARMING ACROSS CHANNEL TO FRANCE," "EXTRA: REDS DECLARE WAR ON JAPAN," "WE DROP ATOMIC BOMB ON JAPS," "JAPS ACCEPT TERMS," "WAR OVER," "YANKS ENTER JAPAN SUNDAY." The last editions marked the end of the war but not the end of my father's stint on the *Greely*. He was still playing music on deck while the ship brought home troops. In the midst of half a million people in Manhattan streets celebrating the war's end, my mother was alone with her letters. After keeping them in New York for many years, did she move them in the bacon box to Vero Beach after my father died to have remnants of their lives close at hand? Had my mother forgotten about them? Were they lost in the tumult of time? I wish I had asked about the letters years earlier. I stored the box of letters in the attic in New York with the rest of the memorabilia and returned to Florida.

Correspondence is enduring. Letters document the expressions, emotions, events of the time—the language of lovers, the details of music shared by a couple, life on a ship. Many thousands of war letters are archived in The National WWII Museum. Senior Curator Kimberly Guise writes,

Correspondence offers the ideal expressions of the moment—the language of the time, jargon and military slang of World War II. They are time capsules in envelopes. The physical pieces themselves offer much in the way of information: the paper they were written on, the handwriting or typeface, the V-mail stationery, the stamps or lack of them, the APO (Army Post Office) addresses, the censors' marks. They may say a great deal about the resources available or the time it took to get a simple message like "Mom, I'm OK" across the globe. Some letters were returned marked "Missing in action" or "Whereabouts unknown." Mail was an invaluable lifeline to those at home. The power that a simple letter had upon the morale of one in service is incomprehensible. In an age when immediate contact was impossible, waiting on the mail was a constant preoccupation. When one did not receive a letter, it was a constant source of worry. The mail itself is a very popular topic of correspondence, nearly every letter either promising to write soon or begging the recipient to write more often.[2]

When sailors my father knew and shared music with did not receive letters, they fretted and worried, the void in communication allowing a host of unwelcome thoughts, anxiety, as well as furtive hopes for the next mail call.

The letters also document life on the home front, where historic changes were taking place in American culture, and couples relied on letters to deal with daily challenges in an emerging world. Guise writes,

Letter writers instruct wives to buy footballs for sons for Christmas or talk of wishing they could be at their formerly despised job back home. They want to make sure their daughter had her shots, that the rent was paid, or that the insurance was taken care of should they not make it home. They tell of foreign shores seen, exotic animals, typhoons, and native populations. Those back home tell of school assignments, blackouts, bond drives, rationing, and prayers for the safe return of loved ones.[3]

I did not attend my mother's memorial service in New York. My brother, Bill, and sister, Penny, took care of arrangements and represented the family. Moving to New York was part of my retirement plan, but I was not able to leave Florida to attend the service in New York. After completing the move a few years later, I had many projects on the agenda, among them cataloging and analyzing the letters. If I found enough substance in them, I would write a book about my parents, music, and the war as portrayed in the letters and records. Several years after depositing the box of letters in the attic in New York, I opened it and was devastated. The letters were in disarray. Some envelopes were empty, some letters had no envelopes or dates, only days of the week. The chronological order that I noted years earlier was lost in the confusion of paper. Again, I closed the box, reluctant to take on the project with a confused or incomplete record. I asked my brother about the letters and learned that he and my sister went through them looking for passages to read at my mother's memorial service. He said he had taken only one letter and I could have it. My sister was less optimistic. She would look for them. I queried for some six or seven years, being assured that the missing letters were around somewhere. The final answer was delivered in 2021: "I must have put them back in the box." They were gone.

The Coast Guard cutter *Northland*, patrolling near Greenland, was the first American unit to make contact with the enemy in the months leading up to World War II. The *Northland* seized the *Buskoe* and later captured Germans manning a weather station.

CHAPTER THREE

⚓

Music of War

For centuries, music has played an important role in wars by inspiring troops, consoling them in their injuries, and boosting morale on the battlefield and at home. Every war had its unique circumstances and unique music to define it. "I don't believe we can have an army without music," Confederate General Robert E. Lee said. The Confederacy adopted "The Bonnie Blue Flag" to replace "The Star-Spangled Banner" as a statement of national identity. Dozens of camp songs inspired Confederate troops. Union troops had their favorite songs. "John Brown's Body" inspired troops by portraying the death of the fiery abolitionist, helping to change the agenda of the Civil War from one of national unity to emancipation. A decade later, the Irish quickstep "Garry Owen" became the official song of General George Armstrong Custer's 7th U.S. Cavalry Regiment as they marched toward Montana and their demise. "Over There" was one of the most inspirational songs of World War I, encouraging young American men to fight in foreign lands. A few decades later, Australian troops marched to the popular song "Waltzing Matilda," composed in 1895 and revived for World War II.

Shostakovich's *Seventh Symphony*, inspired by the Nazi siege of Leningrad, featured musical techniques that reflected militarism and optimism that Russia would prevail despite the devastation of Leningrad. The music endured long after the war as a testament to the Russian spirit. In July 1942, the United States expressed solidarity with Russia as the NBC Symphony Orchestra performed the *Seventh Symphony* in a special radio broadcast. A popular Russian song was "The Soldier's Waltz," featuring the stanza "It's a long time since I've seen my loved one; we are so far from the homeland."

"Lili Marlene" was beloved by German troops in the early days of World War II and was broadcast by Belgrade radio to troops in North Africa.

The theme, separation of a soldier and his girlfriend, was universal in war. English soldiers loved the melody and called for a translation. Joseph Goebbels, propaganda minister for the Third Reich, was disturbed by the song's popularity among the Allies and angered that German singer Lale Andersen, who popularized the song, was associated with a Jewish man. Goebbels sent her to a concentration camp. That action failed to silence the music. American audiences adopted the song after German-American movie star Marlene Dietrich sang it.

Although "Lili Marlene" never topped the charts, it sold steadily as both sheet music and recordings. People worldwide knew the melody, a fact that made "Lili Marlene" possibly the most popular song directly related to World War II, and certainly with its huge Germany-British-American military following, one of the few common denominators to be found in the conflict.[1]

Many English war songs ignored romance, but instead struck a patriotic note, inspiring soldiers and civilians at home to endure the daily assaults and losses of the early phase of World War II. "There will always be an England," "We Must Stick Together," and "He Wears a Pair of Silver Wings" played over radios and in dance halls, eventually crossing the Atlantic with a message for the United States—enter the war.

While Britain waged war against the Nazis, romance dominated American musical taste. Saxophonist and band leader Jimmy Dorsey landed a 1941 hit with "Amapola" ("Pretty Little Poppy"). Together with vocalists Bob Eberly and Helen O'Connell, Dorsey took the 1924 revival to the top of the charts. The lovestruck lyrics and Latin melody give no hint of war in Europe. Americans' taste in music differed as much from that of England's as did its involvement in the war.[2]

The Japanese attack on Pearl Harbor on December 7, 1941, spawned a hodgepodge of patriotic American songs that expressed rage with little tact or subtlety in lyrics or melody. "The Sun Will Soon be Setting on the Land of the Rising Sun," "It's Taps for the Japs," and "We're Gonna Have to Slap the Dirty Little Jap" were among the songs that made brief trajectories on the music charts before nosediving into the sea of oblivion. A few war-themed songs experienced enduring popularity.[3] "Praise the Lord and Pass the Ammunition" was likely the first mainstream popular song about the war. Based on the supposed words of a chaplain on a ship under attack, the authenticity of the story remains in doubt. Along with "Comin' in on a Wing and a Prayer" it became somewhat popular with American audiences in 1943 while dozens of romantic songs void of war themes floated to the top of the charts and lingered after the war.[4] Only twenty-seven war-related songs reached the top of the charts during the entire war, and it was a position they quickly lost.[5]

While Arthur Schnell was studying music at Ithaca College in the late 1930s, big bands had emerged playing popular music derived from swing, a blend of sounds from New Orleans, Chicago, Kansas City, Dixieland, and blues. Swing bands, normally comprised of a dozen or more musicians, tantalized audiences with new sounds, melodic freedom, and percussive beats. Charlie Barnet, Count Basie, Duke Ellington, Benny Goodman, Harry James, and Glenn Miller were among the big names at the time.[6] Art's favorite musician was Tommy Dorsey, whose liquid trombone interpretations were a model for much of his musical aspirations. Although I didn't understand his musical preferences when I was young, in reading his letters and learning about the music of war, I came to appreciate his choice of instrument—the trombone. Glenn Miller, Tommy Dorsey, and many other well-established musicians found the trombone a perfect instrument for big-band dance music. In the Manhattan Beach Coast Guard Marching Band, trombones were front and center of the parade.

Sheet music flew out of presses, records were cut at a frantic pace, disk jockeys, radio shows, movies, and musicals entertained Americans in record numbers. The military recruited top musicians to form and direct military bands. Yet no song emerged as the definitive American World War II song. In fact, after the initial musical rage over the attack on Pearl Harbor, American audiences turned not to songs of war and patriotism but to slushy romantic music and lyrics. The mood of the nation was somber. The United States scored important victories in the Pacific but at a high cost. The Battle of Midway was won at a cost of several warships, hundreds of aircraft and airmen. Six months later, U.S. forces prevailed in the Battle of Guadalcanal with a loss of 1,600 troops in combat and many more from malaria and other tropical diseases, as well as twenty-four warships. Prospects were improving for the Allies, but the end of the war was nowhere in sight.

By the time Art enlisted in the Coast Guard as Musician 3rd Class on June 19, 1943, the sizzle of big bands was simmering.[7] Dance halls, clubs, and restaurants—popular big-band venues—faced war-related restrictions. Food and gasoline rationing discouraged travel to the musical venues, while frequent blackouts and brownouts put a damper on evening performances. A 20-percent tax on entertainment took its toll on commercial music, and the musicians' union attempted to ban its members from recording music. Not all big bands were gone when Harry James's band, featuring singer Helen Forrest, played the 1943 hit "I've Heard that Song Before." The music was ablaze with horns and drums as it embraced the slush theme about a melody that called to mind a love affair. Another hit was "Paper Doll," almost void of horns, piano, drums—a sign that big bands were succumbing to the hardships of war. Many other romantic songs made the charts that year, including "As Time Goes By," "I'll be Home by Christmas," "One for My Baby (and One More for the Road)."[8]

Like many Americans, Art loved the big band sounds, the romantic music, the bold brass of trombones and trumpets, the throaty wailing of saxophones, and the smooth voices of crooners. But the federal Music War Committee did not. The committee was charged with producing war songs that would inspire troops on the battlefield while encouraging Americans at home to produce war materials and buy war bonds. Even with the esteemed musical theater producer Oscar Hammerstein II at the helm, the MWC was sinking almost since its launch.

The sale of bonds was crucial to support the war, and the MWC released songs crafted to inspire people and corporations to invest in the war. "Dig Down Deep" and "Any Bonds Today?" were among the "approved" songs. But when celebrities such as Bing Crosby and Dinah Shore took to the airways in support of bond sales, it was their popular romantic tunes that inspired audiences. Undeterred by popular taste, Army leaders were convinced that listening to patriotic music would make good soldiers, and introduced the *Army Song Book*, containing sixty-seven patriotic songs the troops were expected to memorize. Pages of lyrics were wrapped around rations to make sure they reached the front. Instead of adopting the official songs, soldiers made parodies of the lyrics that satirized military protocol often with profanities. The approved song "Bless 'em all" evolved into a number of variations including the verse "They say there's a Fortress just leaving Calais/Bound for the Limey shore/It's heavily laden with petrified men/ And stiffs who are laid on the floor." General George S. Patton was not dismayed at the parodies and the profanities of his troops. He believed "an army without profanity couldn't fight its way out of a piss-soaked paper bag."[9] Despite the lukewarm reception of "appropriate music," the U.S. Army forged ahead in 1943 with official songs in the *Hit Kit*, a pocket-size songbook that troops carried into battle. The *Hit Kit* was updat-

The pocket size *Hit Kit*, the Army's attempt to infuse troops with inspirational music, failed in a year.

ed monthly with copies in the millions. But troops showed little interest in the *Hit Kit*, and the program lasted only a year.[10] Despite the best efforts of the federal government, sanctioned music made a mere cameo appearance on the musical stage of World War II. Young and Young, authors of *Music of the World War II Era*, wrote,

[N]o one piece of martial music ever emerged as the song associated with World War II. A few earnest efforts were doubtless made by tunesmiths, but neither the civilian public nor the troops fighting the war ever responded to their labors in a positive way. Most war-related music slumped; a song might attract momentary attention, but it usually proved fleeting. Romance in all its forms, the slush derided by those in authority, remained the people's choice, and the most successful compositions in this area seldom referred to the conflict at all.[11]

Advancing technology took control of the musical message from the federal government and sidelined songs of patriotism and the heroics of war. The musical agenda instead was determined by producers in the emerging radio industry, the burgeoning recording industry, sound films, and the American public, which consumed romantic music as never before.

> Rather than dominating the musical offerings with edifying classical music or patriotically promoting the artistic achievements of domestic composers, radio programmers and record distributors largely abandoned efforts to raise national consciousness and instead gave in to the demand for dancing, sentimentality, romance, and fantasy.[12]

Patriotic songs had their place, although clearly second place, in home-front music. Some musicians and publishing houses marketed patriotic songs, either believing there was a lucrative market for such music or to help the war effort by aligning with the federal government's musical taste. Robbins Music Corporation in New York published *Songs for America* in 1941. This collection of songs portrayed the greatness of America, kindling patriotism and boosting morale. The themes of these songs suggest a recipe for patriotic songs to follow to gain access to the American psyche. Some songs portrayed the enemy in a negative light but also as a formidable adversary. This formula avoided creating a false sense of security for Americans and maximized the heroics of soldiers and sailors in battle. Other songs portrayed unity—Americans united in common action, love of freedom, and unity of people from different countries and backgrounds in the fight against a common enemy. An inspiring title was essential to penetrating the intended venues, including "schools, assemblies, music groups, homes, and social groups," as described by Robbins Music Corporation inside the front cover.[13] Everyone was searching for that great American war song. But few if any made the upper reaches of popularity.

The original Coast Guard Band was formed in peacetime, some fifteen years before the start of the Second World War. Based at Coast Guard headquarters in New London, Connecticut, the band played for the commissioning of training vessels, for base morale, and in 1929 for President Hoover. Among the founders was John Philip Sousa, whose rousing marches exuded patriotism with every blast of brass and thunder of drums. "Semper Paratus"—"Always Ready"—became the official Coast Guard anthem. In my preteen years, Art would occasionally sing or play parts of the song with a good deal of pride, sometimes accompanied by a rousing ostinato on the trombone.

> We're always ready for the call,
> We place our trust in Thee.
> Through howling gale and shot and shell,
> To win our victory.
> "Semper Paratus" is our guide,
> Our pledge, our motto, too.
> We're "Always Ready," do or die.
> Aye! Coast Guard, we are for you.

When the war began, the Coast Guard formed regional bands including the Manhattan Beach Training Station Band. The bands had dual musical roles—play the music the federal government sanctioned and play music the public wanted to hear. They had vast responsibilities—keeping up morale on bases, keeping up morale on the home front, entertaining troops in transit, inspiring troops going to war, and inspiring the public to buy war bonds. Under the direction of Rudy Vallée and his signature silky voice, the 11th Naval District Coast Guard Band in Wilmington, California, relied on celebrity musicians such as Tommy Dorsey, Paul Whiteman, Woody Herman, and Fred Waring, as well as boxer Jack Dempsey, comedians Abbott and Costello, and other big names to promote bond sales. Judy Garland, Doris Day, Marlene Dietrich, Dinah Shore, were among the dozens of celebrity vocalists who displayed their talents in the name of war bonds. Other armed forces bands contributed as well. The Manhattan Beach Coast Guard Band, the Army Band, the Military Academy Band, and the Army Air Corps Band were part of a special music division that performed for bond rallies, parades, and recruiting drives.

Art joined the Manhattan Beach Training Station Band in June of 1943, while still in basic training with a full schedule of performances on and off base. The Manhattan Beach band was in demand for off-base functions. The Wednesday radio broadcasts on the Mutual Broadcasting System—popularly known as the Mutual—featured the full Coast Guard band with occasional solo performances. One Sunday in late June, the military band took over the Sunday broadcast that usually featured a dance band. Surely Art and other band members were pleased to sideline the patriotic medleys for the slush and romance of dance music. The bands played for movies, dances, and other social events. In most performances, such as bond rallies in the streets of New York, the Coast Guard Band played popular patriotic marches such as John Philip Sousa's "Stars and Stripes Forever" along with a medley of traditional American favorites.

On one occasion, a five-piece band played at Manhattan Beach sick bay, where an apprentice seaman was recovering from the mumps. The sailor posted a letter of gratitude in *Harpoon*, the base newsletter, and the editors used his gratitude to portray the healing qualities of music.

I really must say, it was swell to hear you play this afternoon. I look from day to day to hear you play. It helps a person a great deal especially when he can't get out or is unable to see or talk to anyone on the outside... Looking forward to your next appearance at this spot. Our undying gratitude.
Postscript: Would you please play "As Time Goes By," "Wait for Me Mary," "Moonlight Becomes You."[14]

Funding the production of World War II was as important as preparing troops for combat. When war broke out in Europe in September 1939, American productivity was sluggish, unable to provide full support for allies. Although the United States was not at war, Roosevelt recognized the importance of industrial production to support the allies. His administration adopted a capitalistic model to meet the demands of war. It was a pricey strategy. Secretary of War Henry L. Stimson described the dilemma. "If you are willing to try to go to war in a capitalist country, you have got to let business make money out of the process, or business won't work."[15]

The federal government's cost-plus-a-fixed-fee program guaranteed the development costs of war conversion, paid a percentage profit on production, and incurred massive federal debt. By war's end, federal outlay reached $98.4 billion, an 11-fold increase over 1939 expenditures. Roosevelt's initial preference was to fund war production with taxes. As war production got underway, he noted, "I would rather pay one hundred percent of taxes now than push the burden of this war onto the shoulders of my grandchildren."[16] But taxes paid only half the war expenditures. The other half was funded by citizens and corporations buying war bonds. And that is why Arthur Schnell and thirty other musicians marched the streets of New York and played music over the Mutual Radio broadcasts. Their contribution to war bond sales helped raise more than $185 billion by the war's end.[17]

"Into the jaws of death," among the most famous of D-Day photographs, was taken by Coast Guard Chief Photographer's Mate Robert F. Sargent from his landing craft at sector Easy Red of Omaha Beach.

CHAPTER FOUR

Whispers of War

Mrs. Frank E. Schoen, commander of the Brooklyn chapter of the Navy Mothers Clubs of America, had a seemingly simple strategy for forming a band at the Manhattan Beach Training Station. She proposed moving the band based on the U.S.S. *Augusta* to the Manhattan Beach base where some 20,000 sailors were stationed. The *Augusta*, a heavy cruiser, had seen action in the early days of the war and was destined to see plenty more. She wrote Admiral Ernest J. King, commander of the U.S. Fleet, in late June of 1942 explaining how the morale-boosting virtues of music would be especially beneficial to sailors on the Manhattan Beach base and noted that her son was a band member on the Augusta. She did not mention that the base would be a secure place for her son and thousands of other sailors to support the war from afar.[1] A week after mailing the letter, the chief of the Military Morale Section advised her that the *Augusta*'s band was not available for transfer and, in any case, would not be needed because a full band had been formed at Manhattan Beach. "Your suggestion is appreciated greatly," the chief wrote, "and I know that your interest in the Service will continue to grow."[2]

The Manhattan Beach Band had some growing to do too. In its formative stage in the late summer of 1942, it was authorized to have fifty musicians but had only ten instruments. Officers debated whether military policy would allow them to compensate the musicians for the use of their instruments. Could they be loaned or leased? If not, the Coast Guard would have to buy them at an estimated cost as high as $7,700. Maintenance costs would be about $40 a year per instrument and would be the same regardless of ownership.[3] The Coast Guard made a deal to have musicians use their own instruments in exchange for covering maintenance costs and replacement if an instrument were damaged in service of the government.[4]

Military bands abounded during the war, based on training bases, foreign bases, war zones, and ships. American musicians aspiring to contribute to the war effort were in demand by all branches of the military. Big-band director Glenn Miller enlisted to direct the Army band and later the Army Air Force band, entertaining troops throughout the world. The Navy recruited clarinetist Artie Shaw to lead a morale-building band, *The Rangers*. The seventeen-piece orchestra performed for troops in disparate locations from Pearl Harbor to Guadalcanal. The Marine Band enlisted violinist and bass player William F. Santelmann as director even before the United States entered the war. Clare Grundman, assistant director of bands at Ohio State University, joined the Coast Guard as chief musician, overseeing ten bands, nine of them regional. The Manhattan Beach Coast Guard Band was a well-established regional band of fifty musicians when Arthur Schnell enlisted. Coast Guard musicians marched and played in parades on the streets of New York in dress uniforms to the cheers of thousands. Art's classification as Musician 3/c was no assurance he would play in any band. His musical performance would determine his role in the Coast Guard band at Manhattan Beach. In basic training, with his sights set on high-profile musical events and radio broadcasts, he carved out practice time with a government trombone. If he failed as a musician, he could be assigned to one of the Coast Guard's wartime roles involving sea duty in naval combat zones where Coast Guard sailors manned ships and landing crafts in beachhead assaults.

Early impressions of life on base, documented in many of the thirty-seven letters to his wife and a few to his mother, reflect Art's hopes of becoming a Coast Guard musician amid the drudgery of preparing for war. Day one of boot camp began with the Coast Guard confiscating a package of goodies his mother had sent. He was allowed to keep the $5 bill. His letter to his wife on the first day reflected the frustration of separation and his fantastical expectation of a reunion. "My darling you can't imagine how much I miss you. After one day it seems that I can't stand three weeks of this till I see my darling again." Up at 5:30, drill practice, march in formation to chow, return to the barracks to clean up and scrub the floor. Following an inspection of barracks, his platoon moved out to the drill field for a grueling hour and a half of marching and exercise. With all the work and exercises, he lost more than twelve pounds. The ensigns in Art's 2nd Company, 1st

Battalion, cracked down on unruly recruits. Eight men were denied liberty, three platoon leaders were reassigned. The ensigns were intent on filling the days with tedious tasks. Every morning after chow, he and fellow platoon members scrubbed and waxed the floors, polished brass, and dusted and cleaned everything for inspection, he wrote to Flo hoping to assuage his guilt at not writing her while he enjoyed the letters she sent. For personal inspection at noon, recruits donned spotless shirts and hats, their shoes polished to perfection. Their uniforms had to be neat and clean, which meant washing them on hands and knees with a bar of soap. Hanging them to dry around their beds between 5 p.m. and 7 a.m. meant sleeping surrounded by a damp curtain. If clothes were not dry by morning, sailors rolled them up wet and stuffed them into the seabags until the next evening.

The platoon leaders seemed unreasonable, prohibiting men from sitting or lying in their bunks until 5:30 p.m. A ninety-second haircut came within a half inch of Art's scalp. The barracks earned the name the "C.G. concentration camp." The schedule of chores was so tight that Art missed a mail call, the high point of each day. His officers did not understand that a musician needed time to practice and prepare for joining the bands.

Art got his first break when Arnold Broido, a fellow musician from Ithaca College School of Music and now prominent in the Coast Guard band, asked him to play for a Mutual Radio broadcast. After his June 26 performance, he wrote Flo the good news: "The band has got some fine musicians, but they do blow loud." He asked her to send his personal trombone as the outlook for musical performances improved. His band leader mentioned privately that soon "I will be off every nite at 4 till 8 the next a.m., with weekend passes, but don't say much about that as it is only granted to musicians." The opportunities he expected as a musician were on the horizon. He advised Flo not to take a job in the Catskills for the summer, as soon he would be off every weekend, and they could share an apartment near base. His musical outlook improved as he became a regular in the Coast Guard band and alerted Flo to listen to the Mutual Radio broadcasts. He would be playing first trombone, the lead instrument in the section.

July 4 was not a day of celebration for sailors in Art's platoon. They were required to scour the barracks, a five-hour task that left the men feeling weary and unpatriotic. But Art's mood lifted when the boys in the band delivered the news that his trombone had arrived. He uncrated it, practiced half an hour, and then watched a show, *Action in the North Atlantic*. If it was the Coast Guard's intention to make war look glamorous, somebody chose the wrong movie. "I guess the fellows didn't feel too pepped up about seeing action after witnessing that movie," he wrote. Most of them would see action.

By early July, Art was itching to meet Flo in the Catskills on his first liberty, but his hopes were thwarted by another itch. "Last night, about

ten men in our company came down with a dose of crabs and I understand we are all going to the doctor for what is known as 'short arm' inspection. Doubt very much that I have them, but it may mean that our whole company may lose the liberty we were hoping for." The next day's letter confirmed it—no liberty until July 1. "Well,

Art Schnell on liberty retreats to his wife's Catskill farm.

sweetheart, I guess we will have to wait another weekend but that won't be too long. I miss you so terribly much though and I would give anything to see my darling for just a little while." The cookies she sent were wonderful but made him feel lonesome. He passed some around to the fellows the previous night just before taps and stuffed the rest into his seabag.

By mid-July, training focused on preparation for war. Physical training involved ten pushups on fingertips, twenty with palms, followed by boxing, wrestling, and jiu-jitsu. In chemical warfare class, the trainees wore gas masks and ran 100 yards while nearly suffocating. His platoon trained to send messages with flags and lights. Art scored 84, 84, and 80, good enough to be a signalman on a warship, a job he hoped to avoid. By the end of the day, he was too tired to attend a Jimmy Durante live show. Instead, he hit the sack and dreamed of liberty with thoughts of training for marine warfare and images from *Action in the North Atlantic* swirling through his mind.

The band began the day at 0800 with colors and ended with colors—raising and lowering the Coast Guard and American flags. Between morning and evening colors, the band played on the drill field and gave Saturday concerts with music ranging from Wagnerian excepts to Gilbert and Sullivan Broadway pieces. Daily morale tours began at 11:45, with the band marching throughout the base, stopping at company headquarters, the canteen, officers' mess hall, the hospital, and other locations where they entertained sailors who labored in ship training, infantry drills, and other work details.[5]

Before the United States entered the war, Congress prepared by enacting Selective Service in 1940, the nation's first draft in the absence of declared war. Before the attack on Pearl Harbor, the nation was deeply divided over involvement in a European war. Isolationists opposed entering the war in Europe and even opposed efforts to provide war materials to European allies while the Roosevelt administration began strengthening the nation's military and tapped the enormous production potential to support the war.

Arthur Schnell registered for Selective Service on October 16, 1940. Six months later, he was classified III-A, a deferment based on dependency, that allowed him to continue serving as a music teacher at Van Hornesville Central School in Upstate New York. After the attack on Pearl Harbor, December 7, 1941, Adolf Hitler joined his Japanese ally and declared war on the United States. Italy quickly followed and the United States was fully involved in World War II. Art was not. He continued to teach music in Van Hornesville until June 19, 1943, when he enlisted in the U.S. Coast Guard as a musician at the Manhattan Beach Coast Guard Training Station.

Before the war, the Coast Guard was largely responsible for port security, ice breaking, navigational operations, and rescue, roles it had played in an earlier version—Revenue Marine, formed in 1790 under the Secretary of the Treasury. It remained there until President Roosevelt brought it under command of the Navy in 1941. By the time Art enlisted, the Coast Guard was fully involved in wartime operations, engaged in deadly combat in the North Atlantic and the Pacific. In the North Atlantic, even before the United States entered the war, German submarines attacked American supply ships and warships. The Coast Guard performed dangerous missions out of Greenland, rescuing survivors from torpedoed ships, destroying German weather and radio stations, and taking prisoners.

During the first half of 1942, Nazi Admiral Dönitz's "Operation Drumbeat" was in full force to disseminate allied shipping in the North Atlantic, where U.S. cargo ships were easy targets. In just two weeks of the first wave of U-boat attacks, Germany claimed twenty merchant ships with 150,000 tons of cargo. Shortly after, the toll was seventy ships. The United States initiated the convoy system, deterring U-boats and reducing losses.

Art enlisted just after the "Bloody Winter" of 1943, when Germany renewed attacks on allied shipping, taking advantage of the most severe winter storms in fifty years. The battles were waged in a gap in air defense south of Greenland in what was to become known as Torpedo Junction, where, at one point, more than 100 U-boats in Wolf Packs hunted U.S. ships. Coast Guard cutters armed with depth charges destroyed U-boats and performed dangerous rescues.

The Coast Guard's dual responsibility led to confusion as it sought to balance naval combat with its traditional role of coastal and port security. The dual responsibilities also created an identity crisis for the Coast Guard as its image seemed to be engulfed by the Navy. "It is just as I feared, that companies and the general public believe that we are no longer the Coast Guard but simply an addition to the Navy," Public Relations Officer Captain Ellis Reed-Hill wrote to the editor of the U.S. Coast Guard Magazine in 1942. "We have got to do our best to maintain before the public the fact that we are a separate entity."[6] After seeing a Camel cigarette ad that included a reference to the Coast Guard, he praised editor Edward Lloyd for his efforts in promoting the Coast Guard's name.

By March 1942, a year before Art joined the Coast Guard, Navy Secretary Frank Knox recognized the dual responsibilities of the Coast Guard and delineated its traditional responsibilities and sea combat. Most Coast Guard bands remained under the command of the Coast Guard while the Navy absorbed Coast Guard personnel for sea service. Stories of Coast Guard combat operations swirled throughout the Manhattan Beach Training Station, a reminder that at any moment orders for transfer could send sailors to sea.

In September 1941 the Coast Guard cutter *Northland* spotted a trawler flying the colors of Norway on Greenland's eastern coast, attempting to set up radio communication to aid U-boats and German shore installations. Norway was then occupied by Germany. The *Northland* captured the

German ship *Buskoe* and seized its crew and equipment. The *Northland* was the first American unit to engage and capture enemy troops. The Coast Guard's Greenland patrol would continue to oversee operations in the North Atlantic for the duration of the war.

A year later, the Coast Guard was engaged in deadly warfare as the cutter *Alexander Hamilton* completed a stormy eastbound Atlantic crossing as an escort. As the cutter neared Reykjavik on January 29, 1942, a German torpedo ripped into the starboard side, exploding between the boiler and engine rooms, rupturing steam pipes. Twenty of the twenty-one sailors in those spaces died. The next day the *Alexander Hamilton* capsized and was deliberately sunk by destroyer gunfire.

Months later, the Coast Guard cutter *Spencer* was engaged in convoy duty in the North Atlantic. She worked side-by-side with U.S. Navy destroyers, Canadian corvettes, and British escort vessels, crossing the North Atlantic sixteen times. Armed with depth charges and machine guns, she rescued survivors of merchant ships torpedoed by U-boats and attacked German submarines. *Spencer* was credited with sinking two submarines during her North Atlantic patrols. While sinking *U-boat 175* and capturing forty-one survivors, twenty-five of the *Spencer*'s men were injured and Radioman Petty Officer Julius Petrella was killed by shrapnel. The battle was recounted in detail in *Harpoon*, the Manhattan Beach Training Station newsletter July 1, 1943. The full-page, first-person account by *Time Magazine* correspondent William Walton included a paragraph about Petty Officer Petrella's fatal injuries and his last words.[7]

Serving in the battle of Guadalcanal on September 27, 1942, Coast Guardsman Douglas Munro, 22, led a small landing party that assisted in the evacuation of 500 Marines pinned down under heavy enemy fire. Munro turned his boat into the enemy fire and shielded retreating marines. He died from gunfire, but his crew, two of them wounded, carried on until the last boat had loaded and cleared the Marines from the beach. He was posthumously awarded a Medal of Honor.

These are among dozens of high-profile stories of Coast Guard combat operations reported in newspapers, discussed on the base, in the barracks, and at mess halls heralding the future of many sailors of the Manhattan Beach Training Station. "I really am very thankful I am getting permanent detail here after my training," Art wrote July 13. "All the fellows want to get something that keeps them here, but practically all of them will be shipped out." The next day the strict base regimen dampened his optimism and hinted that the Coast Guard had assignments other than music in mind for band members. The day's training involved shooting a 45-caliber pistol. The right hand he conditioned to deftly manipulate the trombone slide was charged with taming a pistol that seemed to leap from his hand.

That evening, while perusing a letter from his wife, he was thrilled to learn that she might take a job in Patchogue, not far from the base where they could get an apartment for $35 a month to share after training. He quickly responded to the good news, closing his letter with, "It's time to practice." In a note about the downside of military music, he wrote, "You know there is an instrument I am getting to hate and that is the bugle."

As the first liberty approached on July 15, Art was determined to get to Flo's Catskill home by way of connections of boats, trains, and buses. Hoping to get off the base at 1 p.m., he would make for Times Square, then to Chambers Street Ferry at 3 p.m. where he'd pick up the Erie Railroad leaving at 3 p.m., arriving in Callicoon near Youngsville sometime that day. If he didn't make the train, he'd get there one way or another. "It's an awful mess, I know honey, but it's hard finding anything way down here."

Fleeing to the Catskills for liberty was an escape from the rigors of basic training to pastoral peace. Flo's parents settled part-time in the hamlet of Youngsville on the family farm in the hills of the southern Catskills while maintaining an apartment in Valhalla, just north of the city. Art, who grew up in Rochester, New York, preferred country life, which he found at the Catskill farm. The farmhouse was at the end of a road of red clay, where foot-tall grass in the median separated the ruts. A black, iron monster of a kitchen stove devoured firewood each morning, and throughout the day for cooking and warmth in winter. Glass eyes of deer heads watched with weary indifference over the faded, drab wallpaper that lined each room. The house faced a shop, its untreated wooden siding the color of old oil, and a cavernous barn, empty except for cow stalls, chaff, and piles of rusted machinery. A Model A truck hand-painted silver was parked by the barn, awaiting chores that hadn't been accomplished in decades. Beyond were woods, graceful hills, and trout streams. Just below the farmhouse, a stream they call "the flats" was a perfect place for swimming and fishing, the latter Art performed with little luck.

Liberty with his wife in the Catskills was only a few fleeting days, but he gained resolve to continue his music and face whatever future awaited him. On his return he wrote Flo from the barracks. "Had lots of fun with my honey this weekend," he told her in his affectionate vernacular. "I does love being with my darling and I loves you very much, forever, honey." He learned he was getting an extra $5, bringing his monthly income to $15. No need for her to send money.

A week later, the tranquility that followed liberty waning, disconcerting thoughts pierced the comfortable shelter of the training base. Sitting on rocks above the ocean, he took in the beautiful view, wishing his "honey" was with him to enjoy it. He penned a letter, describing ominous images that crossed his field of vision. "Right now, I see about five big ships going out. I suppose some poor devils are going across," he wrote.

The next few days were no fun staying behind. He spent a week in grueling galley duty, passing out food and cleaning tables. By the time he finished sweeping the floor, it was time for the next meal. By lunchtime he felt dead tired but still needed to practice the trombone. The mindless galley work obscured the march of days, punctuated only by one night of excitement. About thirty SPARS—Coast Guard Women's Reserves—stationed at Hunter College marched past the barracks after a dance he did not attend. He insisted his thoughts of his wife were not diverted by the spectacle he could see from his window. "You should have heard the yelling and whistling that went on. Pretty corny I call it," he reassured Flo. "You should have seen those fellows who went to the dance just barely dragging themselves around the galley today."

Art slipped away from galley duty to play in a musical octet at the Mutual. "It went well. I was nervous in front of a packed auditorium and broadcast besides." Unfortunately, it did not go well enough to exempt him from further galley duty. With no time to eat, he missed dinner but carved out limited practice time, advancing his musical skills. The demands of galley duty and the need to practice his music were in serious conflict. He needed a way out to perfect his music and maintain his positions in the bands. "Tonight I showed one of my bosses at the galley [my] permit to practice after 5:30. He told me that they needed me after that so I had better switch work detail." To Art's relief, his company commander assigned him to barracks duty, which he shared with ten other fellows. The work was easy, and they loafed around two and a half hours before knocking off for the day. He felt sorry for the unfortunate sailors who had to get up at 4 a.m. With more time to dedicate to music, he played colors every night and began marching with the military band.

Dark rumors persisted about the band's future. Flo was in touch with other wives of sailors and heard the rumors. She asked Art about the veracity of the rumor the band would ship out any time now. His complete confidence was reassuring. "As far as that business about being shipped out is concerned, it is a lot of baloney. If we ever did get shipped to another base, it would be to play [on that base] and that is very, very improbable. The only other place we would go is Washington as the official Coast Guard band and that is also very doubtful." To be sure, Art's musical involvement at the base increased weekly and the band was performing on and off base. He was playing the Mutual broadcasts every Wednesday at noon and keeping Flo informed of the music and band's performance.

Art was washing seabags and mopping floors July 30 when he got the best news yet. "Believe it or not, I am playing second trombone with Dick Stabile's dance band at the dance here tonight. One of the dance band trombonists is sick and they ask[ed] me to fill in. Boy am I worried. I haven't even seen the tunes yet, and they are all manuscript arrangements," which are musical scores that identify precise notes and how to play them as opposed to "fake book" arrangements that provide a melody line, basic chords, and lyrics, allowing musicians to improvise. The invitation to play in Dick Stabile's dance band, even as a substitute, would be a big break for an aspiring dance-band musician. Stabile was a flamboyant saxophonist and bandleader who started performing in theater bands at age 15. Beginning in 1926, Stabile worked a decade as a sideman with several orchestras before forming his own dance band in New York. He enjoyed a lengthy engagement at the Lincoln Hotel in New York before going on an extended tour of ballrooms and hotels across the United States. His vocalist, Gracie Barrie, became Mrs. Stabile and fronted the orchestra when her husband was called up for military service in the Coast Guard in 1942.

Art's musical preference was dance music, but by early August he had come to appreciate the skill of marching in formation while playing music. He was the first musician in the middle of three columns of the marching band, where he was responsible for maintaining distance and pace. "I am beginning to appreciate how difficult it is to march and really play tough music at the same time," he wrote. Evenings playing colors were moments of joy in a regimen of drudgery. Although a big improvement over galley duty, barracks duty was tedious but easy and allowed time to practice. Not yet a resident of the coveted band barracks, he slipped in and fell asleep for an afternoon, awakening five hours later. Nobody missed him in his assigned barracks so why was it so hard to get liberty? "This place is getting on my nerves," he wrote. "Why can't we get liberty once in a while. It gets so darn monotonous I think I'll go whacky sometimes—give me Youngsville any day of the week."

More good news followed when he ducked into the band room after playing colors. "I was just starting to write you when Dick Stabile came up and asked me to play with the dance band, so I played until 1 a.m. and then had to get up at 4:30 a.m. About fifteen minutes ago, I was called to come down to the music office. Now they want me to play at the Hotel Ambassador in New York City for the Navy Ball, so I have till 10 a.m. tomorrow off." At the ball, 100 Navy aviation cadets were sworn in. The dance band played from 8 p.m. to midnight. The musicians took ten-minute breaks between sets, and each got a $2 tip and a glass of beer. Back at the barracks the next day, Art learned that although liberty had been delayed, surely by August 12, he would get liberty most weekends. With a $20 paycheck, he was ready to travel and play music. His military musical career had been launched.

Despite the rumors of sea service, musicians who made it into one or more of the bands felt secure in the safety of the Manhattan Beach Coast Guard Training Station. Others sought similar security in what they believed were permanent positions on base. Married men were encouraged to sign up for some of the traditional Coast Guard roles such as dog patrol, a component of coastal protection critical to the nation's security. By early 1942, the FBI was investigating attempted enemy landings on the East Coast. Shortly after the United States entered the war, intelligence reports identified comprehensive plans by Germany to land agents on U.S. coasts. A declassified Coast Guard document describes the incursions and illustrates the Manhattan Beach Training Station's new role in national security.

Shortly before midnight on the thirteenth of June, a German submarine surfaced about five hundred yards off a Long Island beach. Under cover of the fog, a rubber boat was lowered and four men with four large boxes, silently landed on the shore at a point some hundred miles from New York City. Patrolman John C. Cullen of the Amagansett Coast Guard station, which maintained a regular patrol of this area, was on his nightly six-mile easterly patrol when the landing occurred. Surprising one of the group before the party had succeeded in making its way inland, Cullen stopped and questioned the saboteur leader. Two other men in the background, speaking in German, at once aroused his suspicion but being alone he could do little more than cleverly lead the spokesman into self-incriminating statements. Alarmed, the German at first threatened, then bribed Cullen, who feigned a friendly acceptance of the money only to report the incident just as soon as he could get away. The ensuing arrest of the saboteurs proved the effectiveness of well-established beach protection. On 17 June, four more German agents were put ashore from a U-boat on Ponte Vedra Beach, just south of Jacksonville. Both groups brought with them ample funds of American currency and supplies of high explosives, detonators, timing devices, and so on, designed at a special school of saboteurs near Berlin, where these men were trained.[8]

All German agents were arrested. Six were executed, one sentenced to life in prison, and the others sentenced to thirty years. A comprehensive plan involving all branches of the military for coastal protection was developed under the authority of the Coast Guard. In August 1942, dogs were incorporated into the coastal defense system. A month later, horses were added. Soon, 2,000 dogs and 2,991 horses were involved in coastal defense. During Art's basic training, the Manhattan Beach Training Station was aggressively recruiting sailors for beach patrol using dogs. The base offered married sailors the opportunity to train for the new role, which many assumed would assure them of stateside service in Manhattan Beach. Art declined, writing that dog patrol was "a good racket but I'd be better off in the band." He was the only married man in his platoon who did not sign up. Married men were also encouraged to go to platoon school and become permanent platoon leaders at base. Art declined again, trusting in the permanence of the band.

Those who volunteered for dog patrol and those in the band were to learn that permanent Manhattan Beach residency was not guaranteed. By April 1944, the military no longer considered German landings a threat, and beach patrols were dissolved.[9] The Army took over the security of port entrances, and the Coast Guard focused on its traditional coastal security and rescue roles while maintaining naval combat under the command of the Navy. Band members would learn that music did not exempt them from naval warfare.

During Art's second liberty, he spent an idyllic time with Flo, but the visit was framed by difficult travel arrangements. The return bus ride August 23 was dull, with stops at every corner until it got to Monticello and started moving at a reasonable pace. Near Middletown, the bus stopped to let passengers eat and transfer. A woman sitting next to him talked his ear off and smoked, tossing a match onto his duffel bag. By the time he noticed it, it had burned a silver-dollar-size hole in the bag and singed one of his undershirts. At the base, security officials searched his bags and confiscated ginger snaps Flo had made. They failed to find the candy and sandwiches she had packed and didn't notice the burned hole. He got as much satisfaction at having smuggled goodies into the barracks as he did eating them. Wearing his hat into the barracks, he earned a disciplinary report, reacting with a sarcastic "thank you" to the platoon leader and assuring Flo that the punishment would not interfere with his next liberty—just a few extra hours of work in the barracks.

Making the best of the toxic ambiance on base, he informed Flo that the band played "Love of Three Oranges" by Sergei Prokofiev and hinted that when the war is over, he would be suited to fatherhood thanks to the tortuous barracks life. "I love this getting up every night. I'd sure be handy with a baby."

Once a resident of the band barracks—expected date, September 10—he would have less work and more camaraderie with fellow musicians. Camaraderie among musicians is crucial to producing a good band. They learn from one another, inspire one another, and share in the fellowship of musical performances. While Art awaited transfer to the band barracks, he learned that band member Arnold Broido had met with distinguished musicians—trombonist Walter Beeler and trumpeter Craig McHenry, both of Ithaca College, who Art knew well during his college days. They were accompanied by four or five other Ithaca College musicians, who shared stories and talked music. Beeler became conductor of the Ithaca College Concert Band in 1932 and continued in this position through the 1960s. During the war, he was associated with the United States Army Field Band in Washington, D.C. McHenry, longtime dean of the Ithaca College School of Music, was one of the first administrators in the country to recognize the saxophone as a primary performing instrument and hired the first saxophonist for the college band.[10] Art regretted not being among them as he listened with envy to reports of the reunion of his favorite musicians from college. He hoped to soon be among like-minded sailors in the band barracks, but it would not be soon enough. His expected transfer date was pushed back to September 28. He had little to complain about as he learned of sailors packing their belongings into seabags as they prepared to leave the base for sea duty. "Fifty fellows were just put on call and expect to ship out tomorrow or Saturday, so they won't get liberty this week. These fellows were told that they would all get sea duty probably aboard these new destroyer escort vessels."

By September 29, Art was in the band barracks, and Flo had an apartment in Patchogue. His last letters were addressed to Florence Schnell, 63 Thorne Street, Patchogue, Long Island, a short train ride from the base. With basic training behind him, he spent many evenings and weekends with his wife. Life was good, at least for the next couple of years. As the war in Europe dragged to a conclusion, the Manhattan Beach Coast Guard Band continued its schedule of routine performances and special events with a "gala Military Pop Concert" in mid-December 1944 in support of the 6th War Loan, which raised more than $21 billion, $6 billion above the goal set by the Treasury Department.[11] The Manhattan Beach band, glee club, quartets, and soloists played patriotic, traditional, operatic, folk, and religious songs with selections from American, Russian, and Italian composers, as well as traditional musical pieces at concerts in New York, White Plains, and

Brooklyn. The band's rendition of the "Stars Spangled Banner" honored the U.S.S. *Theenim*, an assault cargo ship commissioned December 23, 1944. Band members did not know at the time that these special events would be their last performances in New York.

My father told me about the bond rallies, the Coast Guard band marching along Manhattan streets lined with spectators and cheering patriots, the band playing a medley of inspirational songs and marches by John Philip Sousa and others. In white dress uniforms, music blazing with brass and reed instruments, drums and bass notes in sync with the thump of marching steps, the musicians must have inspired patriotism that the federal government hoped would lead to war-bond sales. Very likely they did. I found a couple of Art's war bond certificates among the memorabilia in the attic. But to some in the crowd, the soaring sounds of music did not mask the awareness of bloody battles, gunfire, and bombs, details that leaked through the Office of War Information's carefully fashioned portrait of war. Surely, some in the crowds felt a degree of resentment when they considered the plight of loved ones in battles on land or sea, or mourned their loss as they watched band members in dress uniforms playing music to cheering crowds. Undoubtedly some servicemen from combat zones looked with distain on the relatively cushy musical aspects of war, regardless of the number of bonds sold. But the fact is, many in the crowds and many others throughout the country were of prime military age and did not serve at all. About a third of the men between the ages of 17 and 35 remained civilians during the war, while 16 million donned uniforms. While many civilians contributed to the war effort by working in production, they did not risk being a casualty.[12] Everyone in uniform was, to a degree, at risk, musicians included.

How this simmering resentment as well as public perception of privilege might have affected the attitudes and behavior of the Manhattan Beach Coast Guard Band members is impossible to assess. But my father suggested it had little effect. As the war dragged on and the bands played on, some musicians developed a sense of privilege, believing their talents were too precious, their role in morale building and bond sales, too important to risk in battle. Born showmen, many were self-assured, cocky, at times arrogant. Prone to joking and horsing around during formal and solemn military rituals, they may have breached the boundaries of decorum. As my father related on more than one occasion, the band was under review by an admiral in early March 1945, and musicians were not showing the respect for the ceremony that an admiral would have expected. The admiral stopped in front of each musician in the front row. When he addressed the man in front of drummer Shelly Manne in the second row, the man in the front row felt a substantial poke with a drumstick, lunged forward and bumped chests with the admiral, who ignored the incident and continue down the line. Not long after that incident, many band members, Art included, got new orders.

If conduct was a factor in the transfer to California, it was not the only factor. In the early days of 1945, numerous sailors from Manhattan Beach were sent to Alameda, their orders noting that "critical shortage does not permit cancellation."[13] The musicians surely wondered in this new assignment if they would play music again during the war or if they would survive this new assignment.

CHAPTER FIVE

West to the Pacific

March 8, 1945, was a day of uncertainty and emotional trauma for Manhattan Beach Coast Guard Band musicians. They received word that key band members faced immediate transfers. Following a hasty reunion with spouses, they were confined on base until time to board a train for San Francisco. Torn from the security of the Manhattan Beach base, the companionship of their families, and the music they enjoyed, they were destined to serve on a ship out of California, bound for hostile waters and dangerous assignments in the Pacific and Indian oceans, with no indication they would continue their role as Coast Guard musicians. With less than a day to be with loved ones, there was no good way to say goodbye.

"The boys are all taking it pretty well, but we all look very pale and a little tired," Arthur Schnell wrote his wife before leaving the barracks. Humor helped deal with uncertainty. While band members waited in line for inoculations, musician Joe Perna horsed around jabbing them with an imaginary needle as he mocked the pharmacist. At least Art was in the company of musicians who supported one another with talk of music and wives. Most musicians said their goodbyes at their wives' apartments. Flo gave Art a goodbye letter to read on the train along with some snacks. "Just can't believe you'll be gone in an hour," she wrote in a letter he opened when the train was underway. "They can't! Sonny. They can't. The last few days have been like a dream, mostly a bad one except that at least we've been together, and you've been such a darling about everything." She expected to see him again for a trackside farewell at the Pennsylvania Railroad station, but Art intended to board the *Golden Arrow* without further farewells. He talked it over with Joe Perna and Frank Boeberitz and decided "it would be better for all concerned not to call our wives before we left or have them meet us at the train." Their previous goodbyes and subsequent letters had to suffice.

A few wives including Lil Ketchel, wife of trumpet player Belmont Ketchel, met their husbands at the station. She was hoping to see Flo and Frank's wife, Genny. She took pictures for absent wives. Without further words, the musicians boarded the train for Chicago and opened letters intended to ease the pain of frantic separation. "I just opened your letter, dearest," Art replied from the train, "and it was wonderful. It really braced me up."

Flo didn't take the departure as well. A few days later, she was so lonely she could have wept. She and her mother would have cherished the last few moments together at the train station. "We are all fit to be tied because you didn't let us know about [leaving] Thursday night. My mother wanted to see you so badly."

Others resisted the abrupt deportation. Musician Jack Coleman concocted a scheme to delay his departure, surely surprising some and angering others, especially French horn player Danny Cowan. Coleman told his commanding officer that his wife was having surgery for cancer with a 50-50 chance of survival and that Danny Cowan agreed to board the train for California in his place. It worked. "To make a long story short, they gave Danny no choice in the matter so without any warning Danny has to go. They were not even going to let him go up to see his wife and baby, but the chaplain finally gave him permission to go home for a few hours." Art doubted the sick-wife story.

The musicians were unaware that the transfer orders were dated five days earlier.[1] The Pennsylvania Railroad's *Golden Arrow* was bound for Chicago, where they would change trains for San Francisco. California was the portal to the expanding war in the Pacific, where the Coast Guard's participation in the battles of Guadalcanal, Iwo Jima, and others were bloody preludes to the planned assault on Okinawa and the invasion of Japan. The Coast Guard, under wartime command of the Navy, was slated to play a role in delivering troops to the Japanese mainland, where casualties were expected to dwarf Allied losses in the Invasion of Normandy, the Battle of the Scheldt, and the Battle of the Bulge.

The separation was more traumatic for Flo than for Art as she confronted the remnants of their lives while enduring the meaningless routine of work and daily chores. She missed seeing him off at the station, but she had given him cookies, chocolates, and gumdrops. She remained behind to continue alone, a life void of new experiences to temper the pain of separation.

Flo and Genny Boeberitz, wife of musician Frank Boeberitz, consoled each other in the absence of their husbands. Flo continued her music teaching, normally an activity that inspired and energized her. But going to work was a dull routine framed by leaving the apartment in the morning and returning to memories in the evening.

"Even though I hate to go to school today, the thought that you love me will keep me going. Without that I would be just nothing. I can't even remember what it was like before I met you, Mister Mousie. Since then life has really had a meaning."

Flo was a consummate musician who loved to teach as much as she loved to perform. But her latest private pupil and instrument only compounded her ennui. "My favorite: Mandolin," she wrote sarcastically and drew a picture of herself holding her nose. My parents were snobs about music.

Images and relics of their lives together haunted her. Ghosts were in the apartment, on trains, and on radio stations that broadcast music Art had played for the Mutual. She packed his personal items as she prepared the apartment for Genny Boeberitz to become her roommate and engaged in mindless tasks to endure his absence. Genny bought wood filler, varnish, paint, and nails to repair a desk. Together with Flo's mother, Minnie, they did the work Art would have done, unaware that their modified household roles and wartime employment mirrored those of millions of women. Riding the local trains to work carried emotional risks. "I sat behind a sailor and a girl who were positively drooling into each other's eye—Geeeeeeee!! Heck!"

She contemplated abandoning the apartment and moving to Valhalla to stay with her family. The hope of a Frisco reunion kept her spirits up as she packed his belongings. "I'm packing all your things & I feel terrible about it but am looking forward to packing for California."

The war in Europe was going well for the Allies in March 1944. Allied victory in the Battle of the Bulge, a month before Art boarded the train, heralded the collapse of Germany. But the outcome of the Pacific war was as uncertain as the band's role in naval warfare. Although stories of the Pacific war dominated the news, neither husband nor wife mentioned anything about battles in their letters, as if to shield each other from fear of torpedo attacks in hostile waters, high-risk missions along beaches of Japanese-held islands, and the assault on the mainland. Doubt seeped into one of her letters.

"I'm going to be awfully lonesome for you, Bunny, & can't wait until I see you again. At least, we'll both have that to look forward to, won't we?"

She counted on his letters to provide the excitement of travel as the train rolled through states and landscapes she had never seen, vicariously enriching her life while she waited for the end of the war. "Once you get going, you enjoy seeing all those new places and writing me all about it. I just can't wait for your letters, Dear Bunny. Now I'll have another batch to tie with a pink ribbon. Or shall I make it Navy blue?"

Art was enjoying the cross-country trip as if there were no war at the end of the line. By the end of the first day, he had traveled farther west than he had ever gone, joining a war migration that displaced some 20 percent of Americans. They moved in all directions, their destinations determined by the demands of war. The massive numbers of troops headed for California were joined by civilians lured by jobs in West Coast war production, where half the nation's shipbuilding and airplane production took place. California grew by two million inhabitants during the war from a steady westward migration mostly by train.[2]

Each time the *Golden Arrow* stopped, Art dropped a letter or two in a Post Office. He and his musician buddies shared some laughs, while he and Frank enjoyed wonderful, semiprivate accommodations in a Pullman sleeping car. The Pullmans were the ultimate in comfort at the time, with upper and lower berths, padded seats, a restroom at each end of the car, and a smooth ride. A porter made the beds each morning while Art and Frank sat comfortably in seating intended for four. The Pullman porters were African Americans, well-trained and well-dressed, many of whom had fled the Jim Crow environment of the South years ago to seek employment. By the early twentieth century, Pullman was the single largest employer of African Americans, almost all of whom worked as porters.

"I'll tell you, darling, these Pullmans are wonderful. I have the upper berth over Frank, and when I finally got used to the train moving, I slept well," he wrote after his first night on the train. In the upper berth, he experienced more swaying than Frank did below. But the Pullman was a smooth ride punctuated by the clack clack clack of the tracks. As the train took the musicians through Pittsburgh and points west, many of them hoped for a Frisco reunion with their wives.

By the time the train reached Indiana, the clacking of the tracks was a monotonous percussion as boring as the flat, drab land he could see through the window. But the food was good—scrambled eggs for breakfast and roast duck for lunch. The best meals were the treats Flo made and packed for the trip. Most married musicians had food gifts their wives hastily prepared or purchased before departure. Band member Harry Brown had four large sticks of salami that he shared.

As the train approached Fort Wayne, Indiana, four merchant marines who had been drinking since New York had become pleasantly polluted, swaying in dreamy rhythm to the rocking of the train. Art understood the motive behind their intoxication—a celebratory practice to wash away thoughts of the destination or the people left behind. For the intoxicated and the sober, the war seemed a world away.

They continued on to Chicago Union Station, one of the busiest in the nation, with connections to most major railways. It was a majestic structure with sweeping limestone exteriors and ornate interiors. The Great Hall, the station's main waiting room, was spanned by a 219-foot-long, barrel-vaulted skylight 115 feet over the room. When the musicians arrived, the skylight ceiling had been blacked out to make the station less visible to enemy aircraft. During the war, the Chicago Union Station handled 300 trains and 100,000 passengers a day, most of them troops.[3] A few months before the *Golden Arrow* arrived with its contingent of military musicians, the station was featured on *The Saturday Evening Post* cover with an illustration by Norman Rockwell. The theme of the painting was the crowd of Christmas shoppers. But look closely and you will see many in uniform, a Pullman porter, and Rockwell himself. Band members got quick showers at the nearby United Service Organizations, then boarded the Union Pacific Railroad's *Challenger* for San Francisco. A glance at the passengers on this train was a reminder that the sense of peace and pleasure Art experienced on the first leg of the trip would not last. "Boy what a load of soldiers and sailors on here, but we are having a good time," he wrote.

Good time for now, but they were bound for war, some to serve on warships, some to man landing crafts, and others to storm beaches that bristle with enemy fire. Some to their deaths. For now, on this train, people were fooling around, laughing, playing cards, enjoying the scenery, the service, and the food. Art played some poker but did not do well. Bridge was his game. It fit his character, then and years later. Never a gambler, never a bluffer, he played his cards with practice and calculations. That was the musician in him. In music, you don't take risks and you don't bluff. You can't fool anybody. Putting mouthpiece to lips, bow to strings, fingers to keyboard is the moment of reckoning. In a dance band playing the fake book, the band director points to you without warning, and you stand for your solo. If you are playing from manuscript, as Art did on short notice with Dick Stabile's dance band, you play every note as written. If you're not ready for the moment, it's too late. You're either good or you're not. If you want to be good, you practice, practice, practice. He tried to get a bridge game going but failed at first. When he finally got a foursome, they played regularly. A few band members unpacked their instruments and got a jam session going in Art's Pullman, and for a spontaneous railroad ensemble, it was not a bad show. An enthusiastic group gathered for the entertainment, spilling into the hallway of the Pullman.

"I am enjoying this trip a great deal and would be very happy if my darling was along and I wouldn't be so lonesome," he wrote March 10. He managed to keep the boys away from the gumdrops Flo had packed but shared some chocolates while squirreling away the rest for private consumption. After Chicago, the trip to San Francisco became more exciting with dramatic landscape, musical camaraderie, and a nefarious incident with some Air Force officers.

"We had a little excitement last night; some doll got on the train in Chicago and was making a play for some flying officers. They pulled four of them out of her sack last nite. When the train stopped in a small town in Wyoming, they took her and a couple Air Force lieutenants off the train. It seems that she was too interested in airplanes to suit the S.P. and M.P. aboard." He reminded her that he was not the sort who would hop into the sack with such a woman or any other than her. He closed his letter with words of devotion and exclusive love.

The *Challenger*'s accommodations were first class, the menu, limited, but the fare, excellent. One evening, the dinner was baked fresh fish a la Creole, garden vegetables, mashed potatoes, hot dinner rolls, ice cream, and milk. Or macaroni with minced chicken instead of fish, and plenty of a la carte choices.

When the train arrived in Cheyenne, Art dropped off a couple of letters. Snow was thick and temperatures dropped well below zero. The flat land he saw in the morning gave way to a desolate landscape with flat-top mesas, jutting rock formations, and a mountainous skyline in the afternoon. Just before dark, stony land and scrub pine scenery became rolling hills with ranches and grazing cattle. That evening, dinner was roast beef, potatoes, and lima beans, with Jell-O for dessert.

As the train neared Ogden, Utah, the scenery was impressive, with enormous mountains and canyons tracing a jagged skyline. "Please don't think I'm cracked but this country is beyond belief. The mountains rise up directly from the tracks to about 2,000 feet, and it is simply beautiful. Honey, we have simply got to drive out here in our new car after the war."

After Ogden, the train struggled up mountains and descended to a flat and boring landscape, the ride, intolerable. The clicking and clacking of the tracks was a hammer on his brain. Art had not experienced anxiety on boats, but the 102-mile rail across Great Salt Lake about ten feet above the water—no waves, no wake—made him nervous. Trains don't float. After reaching land, he saw miles and miles of flatland without a house or a person and regretted complaining about crowds in New York. The landscape no longer interesting, his thoughts turned to unfinished business back home. The couple had portraits made, but the proofs were not ready when the train left for Chicago. Deciding which proofs to make into pictures would be accomplished by mail. Flo mailed the proofs, noting that her favorite had been darkened by proximity to light. Art would have to wait until he arrived in San Francisco to read her letter and see the proofs.

Flo hung her hopes on a Frisco reunion in June, a temporary relief from the pain of separation. But her mind soared ahead to a world without conflict, a permanent fix for couples separated by war. Phrases such as "after the war" and "when the war is over" seeped into the couple's letters. These words and phrases were scarce as if each feared to entertain much less hope for life after war, given the risks and the uncertainty of marine warfare. Flo again hinted of doubt amid words of encouragement. "I hope I can come to Frisco earlier than June. Don't forget to write me that letter. I'm going to be good and not waste any money so we can have all the cute things we want after the war, and we're going to have them, aren't we Sonny?"

The train ride after Ogden seemed endless, the monotony of the metallic metronome rhythm, intolerable. The landscape was bleak and boring. The excitement that a few days earlier gushed from passengers dropped to a dribble. Art expressed doubts about a reunion in San Francisco. "Sorry honey but I can't picture anyone taking the bus all the way out here. It would just about kill them."

CHAPTER SIX

Home-Front Hardships

The train trip across America reflected little of the turmoil that was occurring throughout much of the world as the Allies advanced toward Japan. The suffering and hardship of troops in foreign lands were stories in newspapers, radio broadcasts, battlefront films, and memorial services. While people in the United States limited their lifestyles in many ways, most suffered inconvenience rather than hardship. Crucial consumer goods were rationed, and shortages affected everything from clothing and fashions to dinner menus. Individuals and families were uprooted. Those who moved to areas needing skilled labor found a housing shortage and grim living conditions. As women entered the workplace, hordes of latchkey children returned home from school to empty houses, and few daycare facilities were available to help.

But the war enabled a lifestyle of entertainment, travel, and prosperity unlike any time before. The sale of books skyrocketed. Membership in the Book of the Month Club doubled during the war. Annual sales of paperback books jumped from several hundred thousand to about 10 million by 1943. Sales of comic books increased more than fivefold by the end of the war. Movie attendance increased by a third. Although 4,000 of the 5,700 baseball players such as Ted Williams, Joe DiMaggio, and Peewee Reese from major and minor leagues served in the military, the sport flourished with lesser-known players. President Roosevelt considered baseball a morale builder and encouraged its prominence throughout the war. The All-American Girls Professional Baseball was formed in 1943 and quickly became popular. Americans in record numbers enjoyed vacations, crowding bus and train stations for destinations such as Florida. A flood of federal funds and related jobs put money into the pockets of many Americans.[1]

Flo's letters expressed little concern about the risks of naval warfare in the Pacific, focusing more on the pains of separation and descriptions of her activities during the economic boom of the wartime economy. She shopped for new dresses and a checkered suit, traveled to her family home in Valhalla, worked full time as a teacher, gave private lessons, arranged and directed musical presentations, dined with other musicians' wives, and took care of the household. She and Genny went to dinner in New York's Franklin Square after a glee club performance. She did it all in the absence of a husband and proudly informed him of her activities. Yet, her letters show that separation was harder, sadder, and more heartfelt for her than for him. Art's experience of seeing this vast nation for the first time was more rewarding than experiencing it vicariously while biding in Brooklyn, no matter how convincing his letters were.

Flo held her nose figuratively while teaching mandolin. Art jammed with his band buddies on the train. She did chores such as making the bed while Art and Frank Boeberitz took seats in their private car and watched a porter do it. She cooked dinner after work while Art dined on roast beef and roast duck. She watched a love-struck couple on a train while he joked about four Air Force officers pulled from the sack of a Chicago doll. Flo seemed desperate, while Art was having a pretty good time on the train and arrived in San Francisco in an upbeat state of mind. That was about to change.

CHAPTER SEVEN

San Francisco

When news of the attack on Pearl Harbor December 7, 1941, reached San Francisco, the city and its surroundings were quickly fashioned into a shield against future Japanese attacks. Off-duty personnel were rounded up and put to work, harbor defenses went on high alert. Soldiers constructed defenses, hauled sandbags to vulnerable areas, and strung barbed wire along likely landing sites. Troops manned batteries up and down the coast and nestled into concrete observation posts, where they scanned the horizon for Japanese fleets and planes that would never come. The entire Bay area geared up to fend off an attack on the U.S. mainland.[1]

April 1, 1942, Captain Marc Mitscher broadcast his renowned words, "bound for Tokyo," as he prepared the aircraft carrier *Hornet* to depart San Francisco Bay with Lt. Col. James H. Doolittle and sixteen B-25 bombers and crew. After the bombers crashed in China, Doolittle contemplated a potentially bleak military future from the dung pile that cushioned his landing. However, his failure heralded the future of San Francisco and surrounding cities as the portal to the Pacific war and ultimately to Japan. No longer in defensive mode, the Bay area geared up to transport troops, equipment, weapons, and aircraft to the Pacific front.[2]

Thousands of skilled workers were needed for the massive expansion of the shipbuilding industry. Uprooted families enduring the stress of a war culture came together in a rapid migration. Locals and newcomers, blacks and whites, men and women, young and old, worked relentlessly to support the Pacific war. Military and corporate managers carefully managed the flood of personnel and war materials to the Pacific.

Troops bound for the Pacific islands poured into the Bay area. The large number of troops Art observed on the train for San Francisco was a trickle in the flood of troops heading for the Pacific via San Francisco. By the war's end, nearly 1.65 million Army and Navy personnel had passed under the Golden Gate Bridge on their way to the Pacific front.[3]

The Federal Lanham Act was intended to help resettle some 15 million people throughout the country uprooted by the demand for war production. However, the act failed to provide housing for those who arrived in the Bay area. San Francisco and Oakland became crowded dormitories as residents rented rooms, basements, back porches, and garages. Families shoehorned relatives and strangers into their homes.

Victory in the Pacific depended on ships. Aircraft at that time had insufficient range to operate from mainland bases, and the only way to get to air bases nearer to Japan was to take islands by force through amphibious landings. The Pacific war needed ships, and San Francisco shipyards geared up to produce them in record numbers in record time. More than thirty shipyards and dozens of supporting fabricators joined together to create the world's largest shipbuilding complex with components that sprawled from Napa in the north, Sacramento and Stockton in the east, to San Jose in the south.

Among the many shipbuilders of the Bay area, Kaiser shipyards stood out. The superior productivity of Kaiser shipyards was remarkable considering founder Henry J. Kaiser had never built a ship before 1940. He had a reputation for getting things done, having been the head of one of the consortiums that built the Boulder, Bonneville, and Grand Coulee dams. When war broke out, his company began producing troop transport ships and destroyers as well as the renowned Liberty Ships—mass-produced cargo carriers. By 1943, he oversaw the production of 30 percent of the wartime shipbuilding.[4] By 1945 he and his associates had constructed seven shipyards and delivered 1,490 ships, 747 from Richmond, California, yards alone.[5]

By 1945, with the end of the war in distant sight, shipyards met the demand for transport ships that would bring troops home from Asia and ferry fresh troops to staging areas in preparation for the invasion of Japan. The Kaiser shipyard in Richmond met the demand with the production of the U.S.S. *General A. W. Greely*, whose namesake began his military career as a private in the Civil War. Promoted to Major General in 1906, he commanded military relief efforts following the San Francisco

Earthquake in April of that year. The *Greely* was 522 feet long with a beam of 71 feet. She was armed with four 5-inch guns, four 40-mm machine guns, and sixteen 20-mm machine guns. She was designed to carry 3,823 troops with a crew of 356.[5]

Art Schnell and his musician buddies were unaware as they debarked the *Challenger* in San Francisco that the *Greely* would be their home for the next seven months, taking them and thousands of troops through hostile waters.

The Manhattan Beach Training
Station, where early chapters of
*A Sailor's Song: Lost Love Letters of World
War II* are set, was the first military
organization to racially integrate with
whites and African Americans sharing
living and training facilities.

CHAPTER EIGHT

Gateway to the Pacific War

The musicians' arrival at the Navy base in San Francisco on March 12, 1945, was a disappointing finale to a pleasant cross-country trip. "I wish I could give you some good news but there just isn't any," Art wrote to Flo the following day. "You never saw such a flock of brought-down fellows. Our sea bags have not arrived yet, we have no mattresses, no change of clothing, no nothing." The same day at mail call, Art was proud to be the first member of the band to receive a letter since arriving in San Francisco.

The only indications that they would be playing music were unclaimed instruments lying around. The band members had not met the chief musician and didn't even know if he was in San Francisco. The official activity other than waiting was mustering twice a day for roll call. They spent nights in the barracks playing cards to pass the time, slept on bare springs, and wondered—wondered what kind of ship they would serve on and for how long, where they would sail, how they would play music onboard, whether they would survive the war.

They found plenty to do in San Francisco and Oakland, which offered new dining experiences, music, entertainment, and culture. In Chinatown, Art enjoyed a "real Chinese meal" with wonton soup and chicken Chow Mein with mushrooms. "Honey, this is Chow Mein like you never tasted before in your life. The whole meal only cost me 85 cents." Another night, Art, Frank Boeberitz, George Gialanella, Jack Pervis, Joe Perna, and a few other band members went to Oakland for dinner, followed by a couple of "ancient" movies. The movies were second-rate, perhaps entertaining to the throng of troops waiting for assignment to the Pacific but not to these musicians. They sat through *Topper*, a 1937 comedy that combined ghosts with big-name leading man Cary Grant. The musicians might have been more impressed by the cameo appearance of the famed composer Hoagy

Carmichael if they had recognized him. He was not listed in the credits, but his name surfaced in the dialogue. George, played by Grant, leaves a bar and says, "'Night Hoagy!" and Carmichael replies, "So long, see ya next time." The second film, *The Cowboy and the Lady*, featured Gary Cooper and a theme song nominated for an Academy Award. The musical notoriety was not enough to compensate for a lame plot. The musicians left partway through the film. All in all, it was not a bad night, as the West Coast band leaders, aware of their fellow musicians' anxiety and discomfort, invited a few Manhattan Beach musicians to sleep in their barracks that night. "The boys in the band here let us use their sacks, but tonight since we don't want to overdo a good thing, we will sleep in our own barracks on the bare springs," Art wrote Flo the next day. Surely the musicians, uprooted from their Manhattan Beach band, envied the San Francisco band members, with hit singer Rudy Vallée directing shows and broadcasts with a contingent of comedians, entertainers, and singers. That band remained intact while the Manhattan Beach band was gutted.

Finally, the seabags arrived along with the temporary music chief, Harold Brody, who called for a rehearsal. The band went over a few tunes that sounded pretty bad. When the rehearsal adjourned, musicians went to their barracks for more music practice while other sailors assigned to the ship headed to Treasure Island for gunnery practice, firefighting, and a variety of courses in preparation for war. Art knew the musicians' reprieve from combat training would be brief. They all would be trained for war. As he contemplated going to sea, the separation from his wife weighed on him. "I am so damned lonesome I could scream, and I am only lonesome for you, dearest. I walk around here in a fog, without ambition or anything else. It is hell darling to be so in love and attached to someone the way I am to you and to be separated."

Mail call was the high point of the day when sailors gathered in hopes of word from wives and other loved ones. War letters were the only links between those facing an uncertain and dangerous future and those at home dealing with separation. Mail call was a public event, and everyone knew who received letters and who didn't. Were the boys counting who got letters? Joe Perna knew how many letters Art had received and told him how much he was envied. "The boys are saying I got the only faithful wife because of all my letters," he wrote to Flo. "I know I have the sweetest wife in the world anyhow." Sometimes he placed letters under his pillow after reading them. Some sailors eagerly tore open envelopes and devoured words of love and affection. Others retreated to a private place to read and reread, savoring the words of love in handwriting a lover would recognize. Many left mail call empty-handed, the void in communication fertile ground for sprouting doubt. On March 13, Art and Frank Boeberitz were the only ones who got letters. Art was thrilled but understood the letdown for those fellows who were longing for word from their wives. "What are the other fellows' wives waiting for," he asked. Five days later, the lack of letters from wives took its toll on sailors and band members, especially trumpet player Belmont Ketchel. "Some of the boys are getting quite worried about their wives as they haven't heard a word from them. Ketchel just asked me if you had been to see [his wife] Lillian." On March 24, George Gialanella got nine letters from his girl in New Jersey, all airmailed and postmarked March 17. Joe Perna's wife, May, was devastated that he hadn't gotten any of the letters she had written every day, while she received his letters. The *Greely* wives got together and shared letters, pictures, and stories. May showed her wedding pictures to Flo, who noted, "she looks fine though and just as pretty as ever, tell Joe. Their wedding pictures are absolutely the most beautiful I have ever seen." Within days of the band's departure, the wives began planning trips to California.

Musicians immediately found a friend in Chaplain Hugh Miller, a classic, old-school pastor in a congregation of musical sailors. Round eyeglasses saddled on his nose and hung on his ears by thin, curled side arms. He styled his hair with a touch of military, the sides fading to the scalp before reaching his ears. After mingling with the musicians and singing in quartets and small groups, he became a stalwart advocate for a bunch of uneasy sailors. As soon as he received good news, he shared it with the band members. Miller told Art to expect a promotion to second class and an extra $18 a month in the near future. Already dreaming of life after war, Art wrote Flo that he was pleased because a promotion, if it did materialize, would help build up the nest egg.

After residing in San Francisco for only a few days, band members got vaccinations and a twenty-minute notice to pack their bags and prepare to go to Treasure Island, where the rest of the crew had been training. The U.S.S. *General A.W. Greely* was to be their home for as long as it took to end the war. Communication with loved ones would be slow and censored while sailors were on board.

Lovers are offended that their private words are exposed to censors. Once the love letter is written, the envelope sealed, the words expressing the hardship of separation are gems of feelings violated by the eyes of another, regardless of the motive. Although the censors were looking for information that might compromise security, they read every word, searching for secret messages, for reckless details, for any bit of information that might be useful to enemies, anything suggesting a code they did not understand. Art told Flo that xxx for kisses and hhh for laughter and other such shorthand messages would not get through the censors. Art and Flo, and perhaps thousands of other couples separated by war, sought ways to fool the censors. It's not that secret information encrypted in their letters was crucial to the spouse or even to the enemy. Lovers sought a secret domain of privacy in a sealed envelope that no one else should pierce—including me, as after scanning the letters and finding references to a code, I had no idea what the code was or what it was intended to communicate.

After a few days in San Francisco, Art apologized for not telling Flo about the train departure back in New York. "Sorry I didn't let you know about our leaving later than expected last Thursday night, but it would have about killed me to say goodbye to my dearest twice."

Treasure Island, built by the U.S. Army Corps of Engineers beginning 1936, was named for the gold-laden fill that washed from the Sierras and became the manmade island that would be the site of the 1939 World's Fair. It was converted to a military installation for the war. Art was impressed by its sheer size. He wrote that about 100,000 Navy, Coast Guard, and Marine personnel were stationed on the island. The building that housed his barracks was 1,000 feet long, 300 feet wide, and 200 feet high. Ten such buildings were on the island and many more troops, facilities, and equipment were in San Francisco.

For the time being, band members had nothing to do but practice, rehearse, and gawk at the massive military operation while escaping to San Francisco when they got a brief liberty. Art considered San Francisco "one

of the nicest cities I have ever been in; it is so clean and has lovely buildings."
Flo and the *Greely* wives needed no such enticement to prompt thoughts of
reunions. Plans for a June reunion had already surfaced in their letters.

In the meantime, Art and musician Roger Hartman went to San Francisco
for dinner and shopping. Hartman had a history of feigning medical issues.
While on the Manhattan Beach Training Station, he went to sick bay, certain
he was suffering from a serious kidney ailment. He was quickly discharged
with a clean bill of health. In San Francisco, his strategy for surviving a
torpedo attack was beginning to synthesize when he purchased a quantity
of Hostess Twinkies, sustenance for abandoning ship. He and Art enjoyed
a dinner with so many courses including soup, fruit cocktail, and spaghetti,
that when the filet mignon arrived, they were nearly full, saving just enough
space for ice cream, all for $2.50. Then they enjoyed a musical night out with
the boys. A Coast Guard dance band formerly stationed at Bolling Field Air
Force Base in Washington, D.C., was playing at the Stage Door Canteen.[1]
The club was famous as a morale booster for troops headed to war. Bette
Davis served desserts there, Marlene Dietrich and Lauren Bacall danced,
Red Skelton joked, and Bing Crosby crooned. "Some of our boys knew the
fellows in that outfit so we went down to hear the band. It is a very nice
place, and it just so happened that Eddie Cantor [comedian, dancer, singer,
vaudevillian, actor, and songwriter] was entertaining there so I got his auto-
graph on my canteen ticket."

Art felt a twinge of guilt when telling Flo about the amenities and entertainment he enjoyed in San Francisco. "Don't think that any of us are
particularly happy, darling, but we have to make the best of what we have,
and it shouldn't be too bad." In fact, it was getting even better. Art got
word that while on the ship, the band would be under the direct authority
of Chaplain Hugh Miller, an officer likely to treat them with understanding and compassion. He advised Flo to take note of his new address des-
ignation, Div. T, the morale department, which indicated playing music to
make life tolerable for those heading to or returning from war. The band
continued to rehearse, preparing for the big day—the commissioning of
the *General A. W. Greely*. The music began to sound pretty good, although
the tunes were not Art's favorites. "We only played marches, but the chaplain thought it was wonderful, and he is our boss. So that's what counts."

Not all band members adapted well to the new environment. Frank Boeberitz had become troublesome, and he would be more problematic on a crowded ship. "Had a slight run-in with our boy Frank today. He won't bother with any of the fellows when we have liberty, and I'll be darned if I'm going to tag around after him while he wanders around town. All the boys are down on him, and he is going to be a miserable guy aboard ship unless he wises up. He is so sarcastic. Oh well. Don't mention this to Genny."

Flo didn't, but as Genny had not received a letter from Frank, Flo read her Art's letter, skipping over the part about Frank's behavior. "It's too bad he's like that," she replied. "But maybe he's just miserable away from Genny."

In 1942, the Office of Price Management began a program of rationing certain commodities—sugar, coffee, meat, butter, tires, and gasoline, among them. Consumers received coupon books entitling them to limited consumption of critical commodities. Apparently not realizing that Flo, like millions of war wives, was becoming self-reliant, making decisions traditionally in the men's department, Art reminded her to get a gas coupon card so she could make the 425-mile trip to his mother's home in Rochester. On his end, the "eagle shit," as the sailors dubbed payday, and he just got $23. However, the military food was so bad sailors often ate out. Musician Harry Brown, who packed sticks of salami during the train ride, ordered salami to be sent each week while he was in San Francisco.

With encouraging news from the European front, Art began to think about a future without war, when couples could be together, settle down, plan a future, buy a home, and start a family. But war in the Pacific was far from over, and that is where the *Greely* was likely to take him. "It looks like the war in Europe is just about winding up, thank God. If these darn Japs will only fall apart now," he wrote. Weighing on the minds of the musicians was the risk of death even in a war that was winding down, even in non-combat situations. Just three months earlier, Glenn Miller, famed band director and trombonist as well as one of Art's favorite musicians, received an invitation from General Eisenhower to help celebrate the liberation of Paris with a radio program, performances for troops, and recordings for broadcast back home. Miller was in England directing concerts for returning troops. Driven to perform and frustrated by flight delays due to bad weather, he boarded a small C64 Norseman with his friend Lt. Col. Norman Baessell and a 20-year-old pilot. They did not make it to Paris. Miller's name was not on the flight manifest, so his death was not apparent for several days. With the Battle of the Bulge diverting military attention, releasing the news of his death also was delayed. The wreckage of the plane has never been found. When news of Miller's death was released, surely it tore through the spirits of every military musician and much of the American public. The Manhattan Beach musicians would have learned of Miller's death while in New York, when Art was not writing letters home, so his grief was not recorded.

Teaching music continued to keep Flo busy during their separation. She rehearsed students for a choral presentation and kept Art informed of the developments in her music scene. The work kept her busy but did not relieve the ghost-like sense that she and Art were still together. "I just can't believe you aren't going to come walking up the stairs," she wrote. "I seem to be waiting for something all the time." She and roommate Genny Boeberitz initially get along fine, taking turns cooking and enjoying each other's company. They drove to Flo's family's apartment in Valhalla, her mother and Genny, chattering all the way. Tension grew as the two women tried to run her life, provoking a case of road rage. "Those two got me so mad yesterday trying to ride side saddle, so I stood up to a taxi driver—and called his bluff. First time too. He tried to turn right from the outside lane right ahead of me. I kept right on going & we both stopped about two inches from each other—but he didn't get by."

Art was pleased that she was assertive. "Your letters are so cute when you stick to your guns with the taxi driver," he replied. She wrote to him that she and Genny carried a floor-stand radio up the stairs to the apartment, laughing hysterically like schoolgirls as they struggled with it, nearly dropping it, vinyl records falling from the drawers and tumbling down the stairs. He warned her not to do any heavy lifting, advice she likely ignored considering her new responsibilities. He realized after reading her letter, that she had become capable in his absence, accomplishing tasks on her own. Recognizing her strengths and capabilities did not help him deal with her absence. "The old dew drops kind of get in my eyes because I love you so much and am so lonesome," he wrote, finishing the letter abruptly as a rehearsal was called to prepare for the commissioning of the *Greely*.

As the *Greely Grenadiers* prepared to go to sea, the activities on Manhattan Beach base were of constant interest. Their wives sprinkled letters with news and rumors. The Manhattan Beach base had served as an anchor for their minds, where they met, where they rehearsed, where they performed, where they horsed around and bonded. News that life on the Manhattan Beach base was no picnic for the band members left behind gave them satisfaction. Chief Mulder went on a rampage. Flo got the latest during a stop at Sheepshead Bay on her way home. She heard a couple of sailors yelling at her. Assuming they were drunk, she ran until she heard the name "Schnell." Eli Bublick and Jesse Ralph, two band members not yet sent away, reported on music chief Mulder's woes and their own. "Bublick is still going strong," Flo wrote. "Mulder restricted him one weekend and his two dates felt terrible." Mulder had more serious problems. Assigned to arrange a chorus for the Protestant Easter service, Mulder tapped a number of Catholic musicians who refused to sing in a Protestant service. When he attempted to force their performance, the fellows went to the chaplain, who sided with them. The Protestant ceremony went on without live music, and Mulder failed to complete his orders. That's not all. Flo shared information she got from musician George Bland, who did not initially ship out. Bland told her that Mulder had ruffled some feathers on a trip to Washington. On his return to the base, he tried to lay down the law with the musicians, the remainder of the band being little more than a bugle corp. The musicians continued to flout discipline and protocol, showing little respect for Mulder. After Tom Lithio laughed while Mulder was dressing down the musicians, Mulder expelled him from the room. Before Lithio left, "he turned and made signs of shoveling a high pile [of manure]. Even George [Bland] was amused. Mulder threatened to break their rates," Flo wrote. When Coast Guard Chief Musician Clare Grundman came down from Washington to investigate Mulder's conduct, he feigned illness. Art was grateful for the news and knew the musicians well enough to expect continued conflict and resistance to discipline, especially to Mulder's rule. "Guess they won't have any picnic there from now on," he wrote a couple weeks later. Flo's report that three of the remaining musicians got a reprieve from Mulder's rule in transfers to Cleveland fueled resentment in New York and on the *Greely*. "Nice, eh?" Flo wrote. Art reacted with sarcasm and shared the news with Tom Stokes. "That's really hard on those fellows having to go to Cleveland," Art replied. "It seems the worse you are the better breaks you get. Tom Stokes just about ate his heart out when he heard that because Cleveland is 58 miles from his home."

Flo was candid about bad news from Brooklyn, news she could have concealed. Parking in front of a parked car with the whole street ahead of her, she cut in too quickly and caught the rear fender on the front fender of the parked car. It ripped the fender away from the side of her car. Although the other car suffered only a scratch, she and Genny Boeberitz spent twenty minutes prying the cars loose. "And then we beat it. Two cops went by and never said a word." She cushioned the news by noting that it was the fender that stuck out from Art's previous accident. Genny had been jittery riding with her and now probably wouldn't again. That night, Flo talked in her sleep about the fender bender and shared the story with some *Greely* wives the next day. "Everybody asks me if I'm going to tell you about the fender I crumpled," she wrote. "I guess they think I'm afraid to tell my honey everything. I'm not—imagine being married to somebody you had to be afraid of." As they were considering selling the car—nick-named "Chick"—Art wrote that she must get the fender fixed, and he would send her the name of a collision man who could do the job.

With no trouble from her husband about the crumpled fender, Flo looked forward to a June reunion as she floated a plan for a more enduring arrangement for her and Genny—a move to San Francisco. Art had already warned her that, due to massive immigration, apartments in San Francisco were hard to come by. If the Coast Guard would move their furniture, they might find an unfurnished apartment cheaper than a furnished one. "We could have it fixed up the way we wanted. Maybe we could get four rooms, and Genny could fix up an extra bedroom that way. We can plan, can't we?" she wrote from a local train. Art liked the idea, but he was 3rd class and not entitled to moving expenses. On Frank's account, they could move furniture courtesy of the Coast Guard. The only news that could stall a Frisco reunion was the rumor that the *Greely* might go to New York to sail the Atlantic, transporting troops to and from Europe. She asked Art to validate the rumor, but he did not immediately reply. When he did, he rated it a possibility.

The Code

When Art prepared to board the ship on March 21, he knew his letters would be opened and censored. He wrote Flo about a code but worried the ship's censors might have cut that section from the letter. He mailed a second letter from the Alameda Post Office to avoid the eyes of the censors and to make sure Flo understood the code before the ship embarked. Some seventy-five years later, two code letters turned up, one in the Utica bacon box and one that my brother had. Bill had been aware of the code for some time and gave me the letter that explained it. Art adopted a code developed by Frank Boeberitz to let Flo know his location or destination. A letter beginning with "Dear Darling" would spell the location using the first letter of the second word of each sentence. With perhaps a bit of wishful thinking, he provided the sample coded destination in each code letter, with emphasis on the coded letter:

> Dear Darling
> I have been working very hard. We are playing each day for the troops. The weather has been great, and I am getting a good tan. This afternoon we had a great game of cards with the boys. How is everything going in NYC? I intend to write mom today if we are not too busy.

Hawaii. The date of arrival in the port if known would be indicated by a mention of the birthday of some fictitious family member. Flo acknowledged receipt of the code by writing that she understood the meaning of "Dear Darling." They were ready to fool the censors and share information about location, a knowledge that helped relieve uncertainty and anxiety. After the ship had sailed, Flo expressed frustration and sadness at not knowing where her husband was. When two days had passed without a letter, she wrote, "Where is my wandering boy?" While the *Greely* was at sea and she had not yet received a Dear Darling letter, she wrote, "I can't help but wonder where you are. Golly I wish I knew. It seems like such a lost feeling to wonder and wonder where you are and what you're doing. I never had to do that, remember? You were always so good about everything but it's a different feeling now." A dot on the map was tangible evidence of existence, that somewhere on the globe of war, the lover was alive and capable of writing an encoded letter. Encoded details about location expressed in letters seemed an insignificant breach of security because by the time the letter reached the recipient, the ship was no longer in the same location, and no one but the recipient understood the code. But it was illegal, and the military was serious about guarding classified information, especially about the movement of ships, and found breeches at every level.

The Ship

The *Greely* was so large Art couldn't find his way back to his sleeping quarters, which were temporary until the band's quarters were ready. The *Greely* had five decks and most musicians were on the second level. Sleeping accommodations were crowded, with bunks four high and a foot and a half vertical distance between each one. As the band members learned from the chaplain more about the ship and its commander, they could not have been happier with the arrangements he promised—special accommodations for musicians, a library of 2,000 books, two pianos, a Hammond organ, a sound system throughout the ship, an instrument room, a recreation room, and a commander, George W. Stedman Jr., who by all indicators loved music and recognized its positive effects on morale. "We understand the captain is crazy about having a band, so that is a good break." A news article released just before the commissioning referred to Commander Stedman's appreciation of music. "As the *Greely* puts to sea, its most prized possession will be the beautiful Hammond organ, gift of the San Francisco League of Service Men. The organ will be electronically connected to the ship's audio system so the music will be audible throughout the ship."[2] With Chief Mulder raising hell with the musicians who stayed behind in Manhattan Beach, serving on the *Greely* would be a pretty good gig, except that wives of band members were not aboard. Flo was excited about the opportunity for her husband, but a hint of envy surfaced in a letter. "Your talk about the ship is fascinating. A Hammond organ and large library etc. You should have a great time." She wished to be a stowaway. Or a Red Cross woman.

When terrorists struck the World Trade Center on September 11, 2001, Coast Guard personnel directed the evacuation of more than 500,000 people using hundreds of local ferries, as well as commercial and private crafts in the largest maritime evacuation in less than eight hours, with no loss of life.

CHAPTER NINE

Commander George W. Stedman Jr.

Commander George W. Stedman Jr. sailed big ships long before the start of World War II. His nautical career began when he was 19 as a seaman on luxury liners. He passed his qualifying exam for skipper when he was 21 and assumed his first command of a civilian ship when he was 23. He made nine circumnavigations on the largest ships of the Dollar and Grace Lines, dubbed "floating hotels."[1]

Commander Stedman congratulates one of his crew inducted into the Ancient Order of the Deep.

With the United States' entry into World War II, Stedman left his civilian career for the U.S. Coast Guard, enlisting in St. Augustine, Florida. He performed a tour of duty in New Orleans before transferring to San Francisco, where he was given his first military command. Stedman was named lieutenant commander of the *Etamin*, a supply ship commissioned on May 24, 1943, and assigned to the 7th Fleet Service Force in the Pacific. The *Etamin* soon joined 216 ships in Operation Persecution and Operation Reckless in New Guinea battles, transporting and protecting 80,000 troops, their equipment, and supplies as they conducted amphibious landings at Aitape and Hollandia deep in Japanese territory.

On the night of April 27, 1944, while at anchor in Aitape Road Harbor, New Guinea, crew of the *Etamin* saw off the starboard quarter brilliant flares dropped from high altitude, illuminating the sky, the harbor, and their ship. A plane dived toward the *Etamin*, leveling off at 100 feet, traversing the ship from stern to bow. The crew sounded the alarm and turned out all lights as what they believed to be a second plane unleashed a torpedo bomb that pierced the plating on the starboard side at the number 5 hatch, splaying flames that ignited drums of gasoline. The ship seemed to heave and twist, the shock and vibrations knocking out communications. Steam flooded the engine from a super-heater boiler, severely burning several of the engine-room crew. Escape routes were blocked by steam, leaving only a shaft alley for escape. The shaft alley was the compartment where the propeller shafts led from the engine to the stern, where they exited in a watertight seal. With the engine room ablaze, crew opened the carbon dioxide valves, putting out most of the fire before entering with gas masks to quell the remaining flames with fire extinguishers. The main engine was underwater. With the stern settling fast and uncertainty about the ability of the forward bulkhead in the engine room to hold back water, Commander Stedman attempted to tow the ship into shallow water. When a landing craft failed to move the ship, Stedman gave the order to abandon ship, warning that no one was to jump overboard. Plenty of vessels and lifeboats were standing by to evacuate the crew. The *Etamin* crew and about 150 Army port stevedores were evacuated. The injured were treated in an Army hospital. Despite the risk of more fires, Stedman and his commanding officer remained on the ship that night, joined by three other officers and six enlisted personnel.[2]

The next morning, Stedman and crew surveyed the damage. The forward section of the *Etamin* was intact but the midship was a disaster. Pieces of furniture were scattered about, doors were blown off, various hardware and plumbing fixtures were ripped from their bases, toilets and basins were cracked and broken. Fuel oil from a leaking tank floated on seawater. Hatches 4 and 5, and the engine room were flooded. Because of fumes in the Number 5 hold, crew needed gas masks to inspect it. They found bodies of two Army personnel. One was removed but the other remained, jammed in the wreckage. Stedman noted in his confidential report of May 1, 1944, that the crew could not determine if more bodies were in the wreckage. They found about $45,000 in cash and estimated $1,000 was missing in the mayhem.[3]

Stedman and port officers from the base deemed the ship's presence in the harbor posed a risk and decided to tow her to sea after removing perishable food. The *Etamin* was towed to Cairns, Australia, dispensing supplies to the fleet while under tow. Her engines were not restored, and she served as a floating warehouse until decommissioned in 1946.[4]

The Salt newsletter of March 21 introduced Stedman as commander of his next ship, the *General A.W. Greely*, providing a brief version of the attack on the *Etamin* and picking up the story with Stedman returning home. His crew threw a party in his honor, providing his wife a memento of his service in the Pacific. It was a watch, the band made from parts of the Japanese torpedo bomber that had been shot down after the attack. At the party, she presented the watch to her husband. While Stedman had acquired many possessions from his global travels, this watch was his favorite. In praise of Stedman, *The Salt* noted that the "entire crew was saved without loss of life."[5] The official report as well noted the successful evacuation. "The entire ship's personnel plus approximately 150 Army port battalion (stevedores) were evacuated from the ship without the loss or wounding of a single person."[6]

Stedman did not spend much time on land after the loss of the *Etamin*. On March 21, 1945, Stedman took command of the U.S.S. *General A.W. Greely*. *The Salt* newsletter praised Stedman's ability to manage the ship and lead the crew. "The constant number of men who have gone 'all out' to go aboard the *General Greely* augurs well and gets us off to a fine start. For such would not be the case if these men did not have complete confidence in the abilities of our commanding officer, Commander Stedman, and complete satisfaction with the fair-minded way in which he has long been noted for treating those under his command."[7]

Lt. Commander Richard Tewksbury was named executive officer of the *Greely*, serving directly under Stedman. Tewksbury had spent thirty-three years in the Merchant Marines and the Coast Guard, most of the time at sea, before his assignment on the *Greely*.

The *Greely* had its own band, the *Greely Grenadiers*, comprised mostly of musicians from the Manhattan Beach Coast Guard Band. The band rehearsed intensely for several days to prepare for the commissioning, including a vocal performance by music chief William Schallen. "The first day [of practice] I was so scared I couldn't do anything, but I have been practicing like mad and yesterday it was much better. The band will really be OK after we get going," Art wrote. The *Greely* also had a chorus that rehearsed on the deck while a loudspeaker blared, and rain pelted the singers preparing for the Carnaval Community Sing honoring the *Greely*. The chorus director, a Red Cross volunteer who requested anonymity, was undaunted by the weather and the loudspeaker as she prepared the chorus to perform "Ja-Da," "Irish Lullaby," "Marching Along Together," "Over There," and "Braham's Lullaby."[8]

On March 22, a crowd of military and civilian dignitaries gathered on the deck of the *Greely*. The ceremony began with a salute, the "National Anthem" by the *Greely Grenadiers*, and more salutes. Commander Stedman accepted command of the ship, Chaplain Hugh Miller prayed, and Stedman ordered the executive officer to start the ship's time and set the log. Music and festivities continued after the commissioning, and statements were published praising the ship, its commander, and crew.

Commander Stedman wrote in *The Salt*, the *Greely* newsletter, "Each of you has a vitally important job to do on the *Greely*, and each man's job is just as important as the next fellow's. You all know your stations, you all know your jobs, and I feel very confident that I have your loyalty and your cooperation. As long as I am your commanding officer, I will bend every effort to guide you and your proud ship through rough seas and smooth seas, through thick and thin, and return you safely to your homes and loved ones." Aboard the *Greely*'s first voyage were fifty-eight Red Cross women. Commander Stedman was asked how he liked having women aboard a troop transport. "I like carrying them alright." After a moment's pause, he added, "I would love to bring them all home."[9]

After the *Greely* was commissioned, sailors who were part of the original crew were awarded a plank owner's certificate, an informal and honorary title

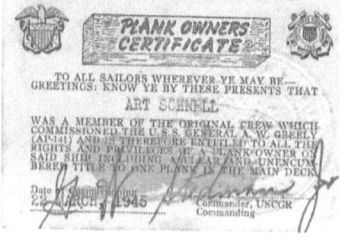

that commemorated the days of "wood ships and iron men." This was the first of several certificates sailors received in the service of the Navy or Coast Guard. Following the commissioning, the *Greely* sailed for the Navy supply depot, where sailors began loading provisions and ammunition.

A headline in *The Salt* proclaimed, "*Greely* First Ship to Have its Own Band Aboard from the Start."[10] The article noted that a few musicians were new to the band. "All other members of the brand-new *Greely Grenadiers* who will throw musical grenades for your entertainment were formerly part of another top-ranking band—the one at Manhattan Beach." The *Greely*'s notoriety as the first ship to have its own band aboard from the start may be strictly accurate, but other military bands accompanied ships en route to hostile territories, as music was considered an essential component of morale and inspiration. One band was the Coast Guard *Invaders* orchestra, a ten-piece dance band assigned to the Coast Guard-manned assault transport the U.S.S. *Samuel Chase*, where the band members manned everything from antiaircraft guns to landing vessels. Between war activities, they performed more than fifty shows in North Africa as well as on the ship, sometimes interrupted by air-raid alerts.

In the fall of 1942, the 40th Army Band of Vermont put to sea on the SS *President Coolidge* bound for the Pacific war. The troop transport ship, a converted luxury liner, was taking reinforcements to Guadalcanal. As the ship approached the island of Espiritu Santo—a military staging port—it hit two mines. Captain Henry Nelson directed the sinking ship to a beach, but it came to rest on a coral reef, where 5,340 troops walked to shore, leaving behind two casualties. Assured that their ship would not sink, the troops left all supplies including musical instruments aboard. When the ship slid off the reef and into deep water, the equipment was lost. After the loss of the musical instruments, a notice appeared in the Vermont newspaper, the *Rutland Herald*, asking for donations of lightly used musical instruments for military musicians.[11]

In the North Atlantic off Iceland, German
U-132 torpedoed the Coast Guard cutter
Alexander Hamilton, killing twenty-six sailors and
wounding fifty-six. She capsized and was sunk
by friendly fire on January, 30, 1942.
The *Alexander Hamilton* was the first American
warship lost to enemy action after the United
States had entered the war.

CHAPTER TEN

Loading the Ship

Between performances on deck to entertain sailors hauling supplies, musicians carried books from the deck of the ship to the library. While they were sorting, cataloging, and putting away a variety of books, Frank Boeberitz and Joe Perna came in with a large suitcase-size box. "They opened it up and it turned out to be a small, portable organ; boy they certainly thought of everything," Art wrote. "As ships go, honey, this is certainly a beauty."

Provisioning the ship continued for about a week after the commissioning. Musicians played two-hour sets to entertain the loading crew, then put down their instruments and carried supplies to the deck. First, they loaded ammunition, enough, Art deemed, to blow him back to Brooklyn. Then they played into the night. All the while tensions built among the loading crew. When several band members were reassigned duties, simmering resentment boiled over. "You know, honey, it's a funny thing everyone worries about the musicians in the service. They don't seem to realize that it takes plenty of preparation to make dance tunes sound good and that it isn't always fun to play. For example, today we were all loading ammunition on deck, some of our fellows were transferred to work on another part of the ship. Immediately all the crew who were loading started yelling 'What! Did those d----d musicians sneak out already?' In the meantime, about ten of the [loading] crew had sneaked away, and nothing was said about that."

Many of the musicians did not take the resentment seriously. Art did but considered it unreasonable. The Coast Guard recruited him as a musician and he would perform to the best of his ability, striving for perfection and overcoming anxiety with each new musical assignment—first trombone, marching middle column, front row while playing; playing dance music with Dick Stabile's band; performing for the Mutual Radio broadcast, accompanying his chief as he sang on the deck of the U.S.S.

General A.W. Greely. He practiced, practiced, practiced, perfecting triple tonguing and other techniques. He improved his embouchure—the union of the mouthpiece and the lips. During rehearsals for the commissioning, he fretted about reaching high notes and determined that playing with dry lips improved his upper-register performance. The band improved as well, following the ragged first rehearsal in San Francisco. Once provisioning was completed, the bands rehearsed five hours a day.

Resentment toward the musicians was brewing beyond the deck of the *Greely.* Musician Harry Brown got a letter from Norbert O'Connor, the bandmaster back in Manhattan Beach, claiming that the musicians likely would experience token sea duty for only a short time. The Coast Guard's strategy, according to O'Connor, was to send some musicians to sea to quell criticism that they were privileged and shielded from hazardous duty. Art doubted they would have brief sea duty and later wrote that the musicians could expect at least six months on the ship. During sea duty, William Schallen would be their musical chief, and most musicians were impressed with his talents. A big-band trombonist and vocalist before the war, Schallen led the Coast Guard's Curtis Bay shipyard band in Maryland before joining the Manhattan Beach band. He had a commanding presence, a tight athletic body, and a rugged, squared-jawed face. He loved to sing and had a great voice, delivering his verses with bravura. Schallen wanted to introduce some complex musical pieces, but the arrangements required more instruments and a bigger band. He asked the skipper for two more trombones and another trumpet. The next day, the rumor circulated that twelve more musicians from the Manhattan Beach band were slated to join the band on the *Greely.* The *Greely Grenadiers* were not happy. Musicians formed a closed community, sometimes fearful of outsiders and protective of their group. They rehearsed and performed as a unit. The integrity of that unit could be threatened by outsiders. Art worried that some of the rumored dozen might be dead wood and detrimental to the band. Other musicians were concerned about how the new band members would be chosen, suspecting that Mulder would send the musicians he did not like. Worse yet, Mulder himself could be among the dozen, given his missteps at the Manhattan Beach base. "The boys are pretty sore out here about it, as we have a good bunch of guys now and we don't want any more except those three [two trombones and a trumpet]. Mulder would last about one day here and disappear. I sure hope we don't have to take those extra men in."

The anxiety over the new band members spawned a flurry of letters between the *Greely Grenadiers*, wives, and band members left behind in Manhattan Beach. Art asked Flo to keep an ear to the ground about who might be transferred. He also told her that if she heard any horror stories about the ship, she should disregard them. Musician George Gialanella had sent a letter to Tom Lithio, who remained in Manhattan, to share with the potential transfers, stating that the *Greely*'s assignment had changed from troop transport to APA—Auxiliary Personnel, Attack transport—a ship that takes troops to beachheads and other war zones. Everyone on the ship was a nervous wreck and trying to concoct some medical excuse for shore duty, Gialanella warned Lithio and the boys in Manhattan, hoping some would find a way to avoid coming to California and serving on the *Greely*. "Those guys will be fudging their drawers," Art wrote.

Schallen suggested that expanding the band could mean a greater chance of getting shore duty. Only a full-size military band would qualify. The *Greely Grenadiers* numbered only twenty-five, not enough for a full band. Resigned to the expansion of the band, Art envisioned the core musicians doing the premier performances without the new fellows while the full band played colors and other menial performances. The new band members would be assigned to cleanup duty and other undesirable chores on the ship while the original *Greely Grenadiers* practiced and played. "We are by no means glad to see them but in what I told you in a previous letter, it will be better in the long run," he wrote.

Confusion reigned as the *Grenadiers* awaited the arrival of the Manhattan Beach musicians. Art grumbled about the quality of the new musicians after learning that Philip Shapiro, Arnold Broido, and Harold Sachs were on their way to California. "Hope the other nine are better than that," he wrote, as he considered the three in the "dead wood" category. He had not yet learned that his old friend George Bland was among the new musicians, as Flo had promised to keep it a secret so that it would be a surprise when Bland arrived in San Francisco. After Bland's arrival, no more musicians were assigned to the *Greely*. The rumor of twelve more musicians was unreliable.[1]

While the *Grenadiers* fretted over the arrival of new musicians, they got word that one band member might be assigned elsewhere, a threat to the camaraderie established by their music. The threat was especially serious as the musician slated for transfer was Danny Cowan, the French horn player who ended up with sea duty because of what appeared to be a prank back in Manhattan, when he was dispatched so briskly that he barely had time to say good-bye to his wife. Danny had been notified that he was reassigned to the U.S.S. *Mayo*, a destroyer sailing in convoys in the Atlantic with a band aboard named the *Invaders*. The band was recruited from the U.S.S. *Samuel Chase* to play for the *Tars and Spars, the Coast Guard Show*, a Broadway musical by Howard Dietz and retired Lt. Vernon Duke. The musical production was intended to entertain and inform the public, as well as enhance recruiting for SPARS, the Coast Guard Women's Reserve. The cast and musicians were made up of personnel from various theaters of war. "The Coast Guard Auxiliary has landed on Broadway," entertainer Ed Sullivan wrote on the playbill. "Never knew how the Coast Guard came up with SPARS (It means "Semper Paratus, Always Ready")."[2] After the show, the band was assigned to the *Mayo* to sail the Atlantic route between the United States and Europe, with Danny Cowan aboard. Musicians planned to lobby the chief to keep Danny Cowan on the *Greely* in hopes Schallen could influence Stedman to pull some strings. They figured Danny had had enough bad breaks.

So had the ship's maintenance crew. The *Greely* hadn't even accomplished a shakedown cruise and things were breaking down. The *Grenadiers* were without reliable electricity and water because of a malfunction of a boiler system. Lights went on and off. Water for drinking and washing was purified from seawater, but the water system was off most of the time, and toilets didn't flush. The shakedown cruise was delayed at least a day. Some crew members took advantage of the last few days on land to escape the rigors of marine protocol. Some were caught sleeping in, shirking duties, failing to show up for galley duty, going ashore without uniform or liberty card, not returning to the ship after liberty, intoxicated, and bruised from fighting, among the many infractions. Most were punished with extra work once the *Greely* was at sea.[3]

Between rehearsals, while waiting for repairs, the musicians had personal provisioning to complete in town, picking up supplies for the chaplain and personal items to pass the time at sea. Art and Danny Cowan volunteered to go to San Francisco, where they joined Chaplain Hugh Miller to buy cases of cigarettes, games, fishing equipment, and other supplies. Art hoped seas would be calm enough to enjoy fishing and not get seasick, an unlikely scenario. The ship had not even weighed anchor when seasickness began to plague some of the crew. When sailors on liberty took the *Liberty* launch to town, the bay was choppy but tolerable. Coming back at night was a different story. "A lot of the guys got stinko and heaved their cookies in that little launch. Ugh!" On return, the musicians were assigned more provisioning — working with the pharmacist's mate to load medical supplies, then working deep in the hold storing canned goods and dehydrated food, "enough on board to feed a city for months."

On Good Friday Art joined a vocal quartet that included Chief William Schallen to sing at a church service. With a second performance planned for Easter Sunday, he hoped to score some points toward a promotion to second class. It was also the Passover, a Jewish holiday. With labor demands high during provisioning, band members got only brief liberties in small groups. On the night of the Passover, Art got liberty with Harry Brown, Danny Cowan, and Harold Brody, who were Jewish. At a Jewish USO, they obtained the address of a family serving a traditional Jewish meal. It began with a few shots of whiskey, then a serving of gefilte fish, chicken soup with dumplings, chicken, matzos, red wine, and fruit salad. "The whole meal was surprisingly good and the hospitality, the best," Art wrote. "Nobody said I wasn't Jewish so guess they took me to be one. Every once in a while, somebody would start talking to me in Hebrew and I would just nod my head in agreement and turn a light purple." Flo was delighted to read the Passover story. "Didn't they make you wear your hat while you ate," she asked.

When the sailors finished dinner and socializing at 11 p.m., they split up. Harold Brody and Harry Brown had booked some female companionship for the evening. Art and Danny Cowan drank a couple beers and returned to the ship, where Frank Boeberitz, now a full-fledged curmudgeon, waited for them. He chastised Art for going out with "those boys" and wondered aloud why anyone would want to go on liberty anyway.

During World War I, Coast Guard servicemen received two Distinguished Service Medals, eight Gold Life-Saving Medals, almost a dozen foreign honors, and nearly fifty Navy Cross Medals. The Coast Guard and the Light House Service lost almost 200 men and five ships.

CHAPTER ELEVEN

Shakedown

April 6, 1945. The U.S.S. *General A.W. Greely* docked at Long Beach at the end of the first day of shakedown—an exercise that prepared ship and sailors for war with battle-station drills, abandon-ship exercises, and music. The band endured a rigorous schedule starting at 0530 with "Reveille" and ending with "Taps" at 2200. The military band rehearsed each day and performed on deck, followed by a two-hour finale by the dance band. Each musician was assigned two life vests; one, a belt that was always worn and inflated with the push of a button. The other was the Mae West type, big and bulky, for emergencies and abandoning ship, dubbed for the generous bust of their namesake. Mae West became the informal name used by the Royal Air Force as early as 1940, continuing throughout the war on both sides of the Atlantic. Sailors readied the movie theater for some 3,000 troops that would soon embark. They stocked the canteen, where Art bought cigarettes for 6 ½ cents a pack. A barber shop was among the amenities the Navy provided for troops. Sleeping arrangements were tight for musicians and tighter for troops. The musicians were in bunks four high on the second deck from the top near the stern. Each band member had a combat role in preparation for war, most of them in damage control.

During library watch dockside, Art wrote two letters to Flo and slipped them to George Gialanella, who mailed them in town beyond the eyes of censors. They described Long Beach and contained the symbols "hhhh" for laughter—a code that would not pass the censors, who had become aggressive as the ship prepared for departure. Other than his attitude toward the censors, Art was happy with his assignment on the ship. "I like it a lot and I am not on deck getting my eardrums broken by those guns."

Executive Officer R.S. Tewksbury wrote in his daily memo that the shakedown was progressing nicely, but sailors could expect hard work to accomplish it in an unusually brief time. The officers seemed pleased by the crew's performance. "So far, everything has progressed satisfactorily," Tewksbury reported, "but it will take a lot of hard work to run shakedown through the limited time available."[1]

The musicians were pleased they were treated well by the officers. "It's so nice to work where you are liked," Art wrote, "and all the officers have commented on how the band boys have pitched in and helped besides doing a good job of playing."

In the library, Art checked out *Stella Dallas*, a novel about a working-class woman whose marriage to a wealthy man fails. Then he started reading *Reveille in Washington: 1860-1865* about politics during the Civil War. He and Flo had listened to a radio review of the latter in Van Hornesville, New York, and reading it on the ship sparked memories of their time together. Later, while shopping with Roger Hartman in Long Beach, he observed the Mexican ambiance of the city, the Spanish-speaking residents, and the flora of the arid climate. Each bought a cactus to send home. That night, Art dreamed of Flo, a dream so vivid "I almost said aloud it was a good one. Now I'm getting cartwheels just thinking about it. Aw heck, cutie." Flo put her cactus in a green dish on plate glass over an egg crate. It was a novelty in New York.

As the separation of war progressed, Art's reflections of a post-war life evolved from sharing a home to having a family. A few weeks prior in San Francisco while waiting to board the ship, he floated the idea of having children after the war. He shared his thoughts with Flo after she had posted that George and Betty Bland, and another couple on the base were expecting a child. "Everyone has them but us, but why, Mousie," he asked. A week later, he refined the family plans to include three children.

Art and two other musicians went to town to find parts for a movie projector that had broken down. Movies on board were important in keeping up morale for thousands of troops and for training. The musicians managed to find the parts in an all-day search that included a visit to Golden Gate Park, "a beautiful place, honey; it certainly makes Prospect Park look like an alley," and to Cliff House, "perched on a mountain with the Golden Gate Bridge on one side and the Pacific Ocean on the other." The search for parts ended with dinner in Chinatown, their last chance for Chinese food.

On April 4, the *Greely* left San Francisco and headed south to Long Beach. Art indicated in his first coded Dear Darling letter that the shakedown was underway.

April 5, 1945. Dear Darling
We all should be in the sack right now as taps has just been blown. I thought I should drop my sweetheart a line, so I wouldn't miss my daily letter. It sure is beautiful the past few days, no rain! There must be something wrong. We expect the other twelve boys any day now, but as yet they haven't put in an appearance. I am getting quite a repertory of popular tunes now, Honey, and have extended my range to a high "C."

It was a rough day for a shakedown. Ship and crew were tested as the *Greely* left port amid strong winds that riled the sea. As the ship plowed big waves into the Pacific, crew plagued by seasickness took battle stations while furniture slid across the floors. Those who had been at sea related the three stages of sea sickness—afraid you are going to die, afraid you won't die, and finally getting well. Only seven band members did not become seriously ill. The sick ones threw up until there was nothing left in their stomachs. Then they gagged and tried to avoid duties. About half of the crew who continued to work were seriously ill. "When I say sick, I mean really sick. Some of the fellows have been so sick they could get up just long enough to 'heave' and then back to the sack. Bublick, Bigatel, Stokes, Ralph, Cowan, and many more are actually green." Tom Stokes was a big guy, lying incapacitated near Art as he wrote letters to his mother and his wife detailing the shakedown. He joked that Stokes would weigh less than his wife, Elsie, when the cruise was over.

Sea sickness was a serious problem for sailors throughout the ages. It is caused by "a conflict in the inner ear, where the human balance mechanism resides, and is caused by a vessel's erratic motion on the water. Inside the cabin of a rocking boat, for example, the inner ear detects changes in both up-and-down and side-to-side acceleration as one's body bobs along with the boat. But, since the cabin moves with the passenger, one's eyes register a relatively stable scene. Agitated by this perceptual incongruity, the brain responds with a cascade of stress-related hormones that can lead to nausea, vomiting, and vertigo."[2] Sea sickness can affect anyone at any time. In 1839, Charles Darwin, who was chronically seasick during the five years on the *Beagle*, assessed the affliction in his book *The Voyage of the Beagle*. "If a person suffer much from sea-sickness, let him weigh it heavily in the balance. I speak from experience: it is no trifling evil which may be cured in a week."[3]

As a young sailor, Admiral Chester Nimitz lost some enthusiasm for the sea along with many lunches during a two-year assignment in the Far East. Sea sickness was a major personnel and medical problem for the Navy and Coast Guard during the war. Between 1942 and 1945, more than 250,000 sailors were admitted to sick bay for sea sickness. They were considered "non-effective" for an average of eight days, meaning they were incapable of performing their duties. More than 10,000 sailors were released from active service as incurably seasick.[4]

"But not me, mom," Art wrote. He proclaimed himself "an old salt" in letters to his wife and to his mother, telling them he had not been ill with sea sickness and even enjoyed going on deck and watching the water when they were not firing the guns. "In fact, I felt wonderful and ate a big meal too." George Bland was a real salt, also, not the least bit seasick. He and Art bonded over their good fortune. Art arranged for the two to be in the same section so they could converse and get liberty together—if there was to be another liberty.

Two miles into the Pacific, the *Greely* reduced way while crew tested the engines. With little motion, the air was still, the sun, brutally hot, and the musicians tanning on deck. Art looked into the deep blue-green water, stunned by the color, the depth, and multitude of fish. Crew were assigned General Quarters, which meant "battle stations." On deck the 5-inch guns were deafening, the percussion waves so powerful that hats seemed to jump off sailors' heads. A Marine on deck suffered a broken eardrum during the firing. Most of the time, Art and other musicians were spared the big gun percussion waves in their damage-control quarters. Art's assignment was fire suppression. In a small room on the second deck near the stern, he would sit in front of the manifold board with numerous valves that controlled ventilation and carbon dioxide for every section of the ship. In the event of a fire and a call from an officer, he would close the ventilation valve to the damaged part of the ship and open the carbon dioxide valve. He practiced this operation while the crew managed fake fires and discharged guns. Between orders, he monitored communication between officers and gun stations and was impressed by the complexity of the ship's protocol. The *Greely* zigzagged, sped up, slowed down, and turned sharply in practice for evading torpedoes. During battle stations, the artillery crew practiced day and night. Below deck, the musicians were assigned menial jobs when not performing or rehearsing. They stamped thousands of life vests with the ship's name and placed one on each berth in preparation for the troops.

Sleeping in heavy seas was a new experience. "It was a queer sensation trying to sleep; you'd feel like you were floating up into space and then you'd start down and think you would never hit the bottom. Guess I am really salty because I enjoyed it, and when we were not working, I was out on deck watching the water. Most of the fellows couldn't even get out of the sack let alone look at the water." The shakedown and sea sickness may have renewed Roger Hartman's efforts to explore strategies for getting out of sea duty. "He claims his kidneys are bad because he sweats so easily and wakes up with pains around his back. We told him it's because he drinks so much," Art wrote in an uncensored letter his friends dropped off at the Post Office.

On April 13, musicians were idly waiting for more work, wondering when the loading would end and the voyage begin, when unexpectantly Art, Roger Hartman, George Bland, and Tom Stokes got brief liberty. They hitchhiked to Long Beach in two separate cars, agreeing to meet at the USO, only to find on arrival that there were eight USOs. They returned to the ship without meeting and were assigned hard labor—carrying large cases of onions, cabbage, and carrots—the fresh vegetables that prevent scurvy. The result of severe vitamin C deficiency, scurvy can cause a number of serious health conditions, including deterioration of the body and mind, and, in severe cases, death. During the Age of Discovery, when ships sailed the Atlantic, Pacific, and Indian oceans in long voyages, scurvy claimed an estimated two million lives. Captain James Cook, during his voyages in the latter half of the 19th Century, realized the cause of scurvy and stocked his holds with fresh fruits and vegetables, reducing the incidence of scurvy among his crew.[5]

During provisioning, musicians noted ominous indicators of their upcoming voyage—duffel bags marked "Australia" sitting in the warehouse awaiting loading.

The shakedown was not solely about testing weapons and machinery, and training sailors for war. The shakedown tested sailors' personalities and the ability to function as a team, to put aside personal conflict. Transport ships provided tight quarters for the crew, and those who demonstrated minor personality issues on land typically exhibited explosive personalities at sea. Before the *Greely* left the dock, Frank Boeberitz revealed his cantankerous side. He became difficult and divisive. After a few days on the ship, he was downright destructive. Trumpeter Belmont Ketchel had it out with Boeberitz in the chow line. "Guess Frank said a few things about 'Ketch' behind his back and they got back to him some way. Honey, Frank is so damned sarcastic and is always giving dirty digs at someone." Things weren't going any better for Flo and Genny Boeberitz in their shared apartment. Conflicts were rising and Flo at times felt uneasy about coming home. One night on the ship, when Art shared stories about Frank's behavior and Flo's account of Genny's conduct, Joe Perna was not surprised one bit. He told Art that Frank had married his match.

Boeberitz's musical abilities had become an issue as well. He was kicked out of the dance band and replaced with one of the new musicians from Manhattan Beach. "He won't practice his horn and... he plays pretty badly, but the way he talks about the rest of the [saxophone] section you'd think he was perfect. Evidently the chief doesn't agree with the opinion Frank has of himself." Only George Bland had anything to do with him, although Bland seemed aware of the personality issues and likely befriended him out of compassion. With one day of shakedown to go, the crew planned for final inspection, when they were required to be perfectly dressed and properly trained. They did their best, considering the laundry system was not up and running, and the crew had been washing their uniforms in buckets. Art's dress blues looked like he'd been sleeping in them for a month.

Young Salt

Arthur Schnell's self-proclaimed status as an "old salt" had its origins in childhood. His toy sailboat, *Gibraltar*, tied for third place in a race on a pond in Rochester when he was in middle school. A year later, *Gibraltar* placed second. A photograph of 1907 portrays a small lateen-rigged sailboat on glassy water with his father, William Schnell, tailor, sailor, and musician at the helm. My dad and his older brother, Bill, undoubtedly got their love of sailing from their father. During the war, Bill was an officer on a supply ship and spent his retirement sailing. Their father also influenced their choice of music as a career. He played violin and performed in movie theaters in Rochester. Art started out on violin but took up the trombone in seventh grade.[6] He studied music at Eastman School of Music in Rochester, completing a preparatory program in trombone before transferring to Ithaca College, where he met my mother. The brothers had a flair for dance music. Bill played saxophone and used the stage name "Billy Swift" for his dance band that performed in New York and Long Island venues after the war, when "Schnell" was not an endearing name because of its German origin. I did not know my paternal grandfather, as he died prematurely when my father was in college. My grandmother, Gertrude Schnell, was a nurse, and in later years, was the personal nurse of George Eastman, founder of Eastman Kodak and Eastman School of Music.

Rumors

In the absence of definitive information about the *Greely*'s destinations, rumors flew fast and furious. The rumor of ferrying troops to Europe had lost its credibility for the time being. Art fended off the worst of the new rumors about naval battles. His mother had expressed concern, apparently prompted by an acquaintance in Rochester, that the 5-inch guns indicated the *Greely* was an attack transport that ferries troops toward beachheads. Citing information from an officer, Art replied that the *Greely* would not stick its neck into danger with thousands of troops aboard. The big guns were for use in the event the ship was "trapped," his euphemism for "attacked."

Flo sifted through rumors at home with the war wives. She spent an evening at Lil Ketchel's, a get-together for some wives of the *Greely* musicians. Lil served coffee, sandwiches, and cake, while the girls draped themselves over furniture for a gab session. "Honestly I haven't enjoyed myself so much since you left. There were nine of us and we all gabbed so much and so fast that we hardly had time to come up for air." They became known as the *"Greely* wives." Flo, Genny Boeberitz, May Perna (Joe's wife), Natalie Cowan (Danny's wife), Chris Purves (Jack's wife), Flip Manne (Shelly's wife), Lucille Broido (Arnold's wife), and Miriam Sachs (Harold's wife), agreed to meet regularly for moral support and to take their minds off the war and the uncertain future of the *Greely*. At a later gathering, May Perna served the wives meatballs and spaghetti, with some cherry wine that, by Flo's standards, was pretty strong, strong enough to inspire Natalie Cowan to demonstrate standing on her hands after a glass and a half. "She's really alright, honey. I think we would have enjoyed their company. I really get a kick out of her." At various gatherings, when they shared rumors about the *Greely's* destination, no one mentioned "India." Flo learned of that potential destination from Art and responded with "Gulp." She closed the letter, "Please don't go to India."

The most reliable and respected source of information was Chief Musician Clare Grundman. Before the war, Grundman, whose instrument was clarinet and later saxophone, was a prolific composer and arranger of music. He served as assistant director of bands at Ohio State University, where he taught orchestration and woodwinds. When World War II began, Grundman enlisted in the Coast Guard, overseeing the main Coast Guard band and regional bands. One of the *Greely Grenadiers* met with him in San Francisco and reported some good news. Grundman said as soon as the band members got a couple of campaign bars, they'd be back in New York playing in the Manhattan Beach band. Campaign bars were given in recognition of service in combat zones. The idea behind Grundman's report was that the Navy or Coast Guard felt compelled to assign the Manhattan Beach band members to combat roles to blunt criticism of privilege and favoritism.

With departure imminent, Art advised Flo not to waste airmail stamps on her letters as they would arrive by ship wherever the *Greely* happened to be. The use of V-mail would be cheaper and faster. The Navy was encouraging the use of V-mail, a system of transmitting letters using photographic copies with the address and the letter content on an open sheet of paper. Only one insect-eaten V-mail letter was among the letters preserved in the bacon box.

President Roosevelt

March 1, 1945, a few days before the musicians arrived in California, President Roosevelt gave his last address over the radio as he reported to Congress on his Yalta meeting with Winston Churchill and Joseph Stalin. His optimism that Germany would soon be defeated inspired a nation to huddle around the radio in anticipation of word on V-E Day—victory in Europe. Instead, a month after the broadcast, radios spilled the news that Roosevelt had died. His death on April 12 in Warm Springs, Georgia, while sitting for a portrait, stunned a nation eager for good news on the European war. Art's letter to Flo written the day of Roosevelt's death conveyed his grief. "The very bad news of President Roosevelt's death just came over the radio in our library, it is still a shock to all of us. If anyone thinks for one second that the boys in the service didn't like him or have faith in him, you should see their faces now. Fellows who two months before were laughing and calling each other every din of a name you can think of, just stood in the library listening to the radio, completely stunned and many of them including myself with tears in their eyes. I am still convinced he was the greatest president we have ever had, and his loss will be felt more deeply as time goes on. Thank God he lived long enough to get us through the worst of this war."

The *Greely* was in port at the time of Roosevelt's death. Five hundred Army officers had just boarded followed by the first 900 troops, the remainder to board a day later. Some 200 Canadian troops joined the large contingent of American forces aboard the ship. "Boy honey they are a scared lot of guys, and you can't help but feel a little sorry for them. Some look quite old but the rest are real young." Battle ready, they knew they were not bound for Hawaii. With a full contingent of fresh troops and officers, musicians and crew, tons of mail, and a mascot—a puppy named "Snafu"—the *Greely* was set to go somewhere.[7] Among the provisions was $100,000 cash.[8]

Art did not get a letter at mail call and realized that only a few days were left when he could expect regular mail. He was lonesome and needed just one letter to get through the night. "Sweetheart I am so lonesome for you tonight I just don't want to do nothing but sit around and mope. Just think darling it has been over five weeks since I left you and each seems like a year." He pleaded for more love letters, and he felt sorry for Angelo Bigatel, who hadn't gotten even one letter and likely wouldn't until the *Greely* reached her next port. With departure imminent, Art signed up his wife for free health care at any Naval hospital. All she needed to do was sign the form he intended to enclose. He forgot to enclose it and sent it with the next day's mail.

As the ship filled with troops, the bands absorbed the new musicians and rehearsed, Art began to appreciate the new members for their music and their camaraderie. He was especially pleased to be playing with Eli Bublick, whose musical talents he admired and whose proximity he tolerated. "Who do you think sleeps right alongside me—why none other than the great trombonist Eli Bublick. The worst of it is that I woke up about three times last nite with a hand in my face so you can see how close his sack is to mine. Never mind, cutie, but aren't you a little jealous?"

Once the troops were settled in, they received booklets, their titles providing the clearest and most ominous indication of the *Greely*'s destination. Each person received two booklets on India, *A Guidebook on Calcutta, Agra, Karachi and Bombay* and *A Pocket Guide to India*, the two confirming the *Greely*'s port destination. *A Pocket Guide to Burma* suggested an ultimate troop assignment in the jungle, where Merrill's Marauders were wreaking havoc on Japanese troops while taking heavy casualties. *A Pocket Guide to China* suggested assignments beyond the Burma theater. *Japanese: A Language Guide* could be useful in the invasion of Japan, if captured by or capturing enemies. Just in case the ship was attacked on the way, the Navy provided the pamphlet *How to Abandon Ship*. It is unclear when these and other booklets were distributed.

With departure imminent, some crew were intent on enjoying their last days on American soil. Some were caught off ship without uniforms or liberty cards, some failed to report for duty in the galley, others overslept, and one was three days AWOL. Others were charged with intoxication, disorderly conduct, and barroom fighting.[9]

The *Greely* left San Pedro, California, April 16, sailing west into the Pacific. Two days later, famed war correspondent Ernie Pyle was killed by enemy fire on the island of Ie Shima near Okinawa. President Harry S. Truman spoke of how Pyle "told the story of the American fighting man as the American fighting men wanted it told."[10] The ship's informal newsletter decried the loss of the president and the journalist.

On her maiden voyage, the U.S.S. *Greely* put to sea with the flag of the United States flying at half-mast. Fresh were the memories of scenes at the staging area where in stunned silence small groups listened tensely to the first radio flashes. Slowly the realization came that upon the dawn of victory it was sundown for a great leader. Darkness settled in the hearts of millions. Franklin Delano Roosevelt was dead. Unusual in the annal of the sea, the *Greely* sailed under the flag of mourning for thirty days of her long outward-bound voyage. The flag flashed in the cold breezes of late spring, in the light of the full moon, in the equatorial air. Late in the second week out it also symbolized a new loss to every American and a special one to millions of fighting G.I.s. The keenest interpreter of the fighting man's hopes and struggles had died... Like many of the men he wrote about ... on the field of battle. Ernie Pyle too was dead.

On Sunday morning May 13, the captain logged this cryptic statement. "The flag was hoisted to full mast." Two men would never see it. Others would live to keep it there.[11]

In Manhattan a few weeks after Roosevelt's death, Flo sang the "Stars Spangled Banner" and directed the Monroe Street School Chorus of fifth and sixth graders in a musical tribute to the late president. On the *Greely*, the radio broadcast news that troops, officers, sailors, and musicians wanted to hear—the end of organized German resistance was expected within the next few days. "The war news looks good, and I am quite confident it will be over soon in Japan," Art reassured Flo. The news renewed rumors and hopes of service in the Atlantic, where the *Greely* could be assigned to bring troops home from Europe after completing its cruise to the Pacific, rekindling hopes among the *Greely* wives that the ship would be based in New York. Despite seeing the duffel bags marked "Australia," Art continued to speculate that the ship would sail for Hawaii and return to New York to sail the Atlantic. Other musicians began to doubt Clare Grundman's word that sea duty for musicians would be brief. They prepared for an extended engagement in the Pacific war.

Band members were beginning to appreciate sea duty as they were assured they'd continue to be musicians while assuming combat roles on the ship, that their talents were respected by their commander, and that they would regularly play for a large *Greely* audience. With a few exceptions, the *Greely Grenadiers* continued as the core of the band they were in Manhattan Beach. But the role of their music in the war had evolved. Before they boarded the train at Penn Station, their primary role was to inspire people to buy war bonds, dig deep into their pockets, and make a financial sacrifice to pay the bills for the troops, the weapons, and the supplies of war. The dance band honored officers' special events and commemorated the commissioning of ships. Radio broadcasts were comforting and inspiring to people with loved ones abroad who needed optimism to continue their support at home. Once on the ship, the band would play for troops about to enter the war for the first time. Fresh troops heading for war spent monotonous hours of inactivity, their leisure time, a petri dish for the growth of infectious anxiety borne by sailing blind to an undisclosed battlefield with an unknown outcome. Operation Overlord—the invasion of Normandy—was almost a year behind them, its stories of victory and casualties familiar to each. Were the troops destined for the invasion of Japan? The troop buildup they were joining foreshadowed the next step toward ending the war, the ultimate destination, the Japanese mainland. Music was intended to take their minds off their fears and concerns, to keep their spirits up, and to inspire their patriotism so they would be strong when tested in battle.

On its return, the *Greely* would carry thousands of troops that had faced and survived the worst of World War II on Japanese islands and in the jungles of Burma, some injured or maimed, some with minds scarred by horrors of war, by the loss of friends. All had suffered in one way or another. But they had survived, separated from families in a realm unimaginable to those who never experienced it. They would return to their homeland to a lifestyle where they would be safe but never in the same mindset as when they left. For these troops, music would be an anodyne, the best the military could come up with until they arrived home, perhaps the only one they would get.

At home, Flo and other teachers basked in good news of the war, but the pain of separation was just a breath away. They dined in Hempstead where they gabbed and joked until Flo smelled gardenias at a nearby table. The fragrance swept her helplessly into the past. "And the music began to play—and it seemed just as if I were at a dance in Ithaca only my Sonny wasn't there. Honest, Bunny, I had tears in my eyes. I just felt so down."

The Coast Guard manned
ninety-nine vessels in the assault
on Normandy, losing eighteen
guardsmen the first day with another
thirty-eight wounded. The Coast
Guard lost more vessels that day
in Operation Neptune than on any
other day in its history.

CHAPTER TWELVE

Somewhere in the Pacific

Commander Stedman's experience as a cruise ship captain served him well as the U.S.S. *Greely* sailed through Pacific waters on its maiden voyage. He knew how to make passengers feel comfortable and entertained. They enjoyed good food, movies at night, card games, and live music by the *Greely*'s military band and the dance band. Each band rehearsed two hours a day on deck to a crowd of idle troops and then performed twice a day to a similar crowd with music piped throughout the ship. The military band performed in the afternoons while the dance band performed in the evenings. Often the Pacific Ocean was calm, and the war, a distant unknown at the end of a pleasant ocean cruise. Between rehearsals and performances, Art wrote a no-news letter to his mother. "Dearest Mom: Well here we are out at sea and all I can say about it is, that there is an awful lot of it." He enjoyed watching the ship plow through gentle swell or looking into the deep blue, so saturated with color it seemed someone had dumped dye into the sea. At times it appeared almost purple. The sun was scorching, and he and most other sailors quickly burned red as lobsters. At night, musicians lay on deck, their inflated life belts under their heads for pillows, the night air, a respite from the heat of the day and swelter below. The deck was a stage for spectacular night shows in a vast sky ringed by water. "It was very beautiful up there looking at the stars, but it made me awful lonesome for my cute honey," he wrote in the only surviving V-mail in the collection of letters.

With 2,923 officers, troops, and civilians on board, Art's main gripe was long lines—lines for chow, lines for the movies, lines to get into the canteen. He skipped the movies in the crowded theater to contemplate a fantasy future that was budding as the end of the war seemed in sight. He longed for the simple country lifestyle that he and Flo enjoyed in the

Catskills or in the hamlet of Van Hornesville, where he had taught music. Despite the crowds, life was good aboard the *Greely*, better in some ways than the daily grind back in Manhattan Beach. The sailors were well treated and didn't even have to do laundry. It was picked up and returned each Friday, clean, dry, and folded. The new boys in the military band turned out

Musician George Bland's cartoon "Jammin' on Hatch 5" published in the *Greely's* newsletter depicting activities on the ship as she sailed through the Pacific Ocean. *The Salt*

to be mostly good musicians.

Musicians and crew worked their talents in idle hours. George Bland was an accomplished artist who quickly became one of the ship's cartoonists, publishing his illustrations in the *Greely's* newsletter. One cartoon entitled "Jammin' on Hatch 5" portrays social life on the *Greely*, with the band on both sides and center. "Property of Chaplain's Office" is scrawled across the image of the band, and Shelly Manne's name is prominent on the bass drum. Above the band, a sailor is making out with one of the Red Cross women. The copy Art sent to Flo June 8 was annotated with *Greely Grenadiers'* names, identifying some of the caricatures. Another cartoon by Bland features a caricature of Shelly Manne, captioned, "You can name the character above.

It's not a voodoo witch doctor and the watermelon is missing so it can't depict 'boy eating watermelon'—despite those pearly white teeth. In reality the skin doctor is none other than Shelly Manne, ace drummer with the *Greely Grenadiers* who has been beating the skins with topflight bands ever since high school days."[1] "*Greely* Summer Cruise—April, May 1945" portrays a medley of activities on the *Greely*, with pop guns fore and aft, sailors dancing with Red Cross women, and a chow line that extends off the ship and into the ocean. (Images are in the Appendix.)

Bland's artistic talents had their origin in his ancestry, a fact he apparently did not mention to Art. I was to learn the origin when I came upon a story of a missing masterpiece as I wrote the final chapter. His grandfather was Henry McArdle, a famous Texas artist. Bland also entertained with his antics, clowning around and getting laughs out of the crew. Some musicians endured isolation amid the crowd. Angelo Bigatel went to sea on a disappointing note without a letter from his wife. He seemed to retreat into solitude. Jesse Ralph was becoming a troublesome musician. He regularly yelled at fellow trombonists Art Schnell and Eli Bublick about their playing. The pair annoyed him by saluting and repeating "Aye, aye captain." Art got along well with Eli Bublick, treasuring his friendship, admiring his musical talent, and enjoying bedtime stories. "You remember me telling you about Eli and his stories," he wrote Flo. "I hear one every nite before I go to sleep." Danny Cowan dodged the transfer order and remained with the *Greely Grenadiers* until the end of the war. It is unclear what tactics Commander Stedman employed to override the transfer order.

In his first letter while underway, Art ran afoul of the censors. When he referred to a big upcoming production, the scissors struck. The letter picks up with "you know, getting the head shaved." What is left out would be obvious to any seafaring fellow aware of the ship's location. The *Greely* was approaching the equator and a longstanding ritual—the Ancient Order of the Deep—was about to take place.

The ritual celebrated the transformation of pollywogs—those who had not crossed the equator—to shellbacks, those who had. In the ceremony, King Neptune, portrayed by one of the ship's officers, presided over a court at Hatch 5, with Davey Jones standing by to exact punishment. First

A. R. Schnell Mus 3/c 'T)
U. S. S. Gen. A. W. Greely
A. P. 141

April 18, 1945

Place: Somewhere in
the Pacific—10 P.M.

My Precious Darling:

Just finished a nice game of poker in which I dropped exactly one dollar and now dripping with sweat I sit down to write my darling. Honey do you remember my writing you complaining about the cold? now just change that to the heat.

We were up on deck tonite for awhile and it was beautiful out, what a trip this would be if it were not for the war and you were with me.

Well honey very shortly we will be going through that big production

—. You know getting the head half shaved etc. Now we are known as polywogs after crossing our name is changed to 'shellbacks'. I'll probably look a mess when they finish with me.

A portion of this letter was cut because it referred to the Ancient Order of the Deep ceremony, indicating the *Greely* had crossed the Equator.

pollywogs in the initiation were crew and officers, then Red Cross women, then musicians.

All pollywogs were summoned before the court to answer charges for crimes against the Ancient Order of the Deep. Typical charges included trying to alter course to avoid crossing the equator, stealing eggs from the crow's nest, and defrauding the royal barbers by hiding during haircut time. Art's charges resulted from a conversation with a staffer of the ship's newsletter. In April, thinking it was May, he mentioned his

A sailor has his head shaved and anointed with oil in the Ancient Order of the Deep initiation.

upcoming June anniversary. The mistake surfaced during the Ancient Order of the Deep when he was charged with forgetting the month of the couple's wedding anniversary.

The musicians wearing only bathing trunks were the last to walk in single file through about fifteen officers swinging paddles at them. "Next we faced the Royal Barber who did a very fast but nasty job on my poor head. Then they covered us with heavy black grease from head to foot besides squirting various assortments of oil into our faces. After being given a couple of invigorating electric shocks, we had to climb or rather jump into a tank of salt water, and as we were climbing out, our 'fannies' were again soundly punished." Some endured a smearing with eggs that had rotted in the ship's hold. The haircuts were so bad some newly anointed shellbacks shaved their heads. Art's hair was a ragged mess with a two-inch wide swath of scalp from forehead to neck. His hazing experience when he joined the music fraternity Phi Mu Alpha Sinfonia at Ithaca College prepared him for abuse and was more severe than the Navy's ceremony, although it did not include the haircut.

Chaplain Hugh Miller takes his punishment in the Ancient Order of the Deep initiation. *The Salt*

Chaplain Hugh Miller, charged with an unspecified crime, endured the most severe punishment. He was put into stocks, commemorated by a cartoon in *The Salt* newsletter. The finale of the Ancient Order of the Deep ceremony was punctuated by water fights and the awarding of certificates, which were not official Navy documents and varied from ship to ship. The origin of the ceremony is unclear. A cartoon in *The Salt* commemorated the ceremony representing the steps required to become a shellback. (see Appendix)

Sailing southwest, the *Greely* encountered increasingly hot weather. Sailors and passengers drank gallons of water each day and took salt tablets to aid in water retention. Art cut the sleeves off a favorite blue shirt and advised Flo not to send perishables. Chocolate melted into a sticky mess. He looked for shade on deck and watched flying fish sailing above the waves for up to sixty feet. Shelly Manne found a twelve-inch flying fish on the second deck. Sleeping on deck at night watching the stars reminded Art of a time when he and Flo were together at Ithaca College. "I couldn't help but remember the time I skipped rehearsal at college, and we sat out in the backyard of SAI [Sigma Alpha Iota International music fraternity] and talked about everything under the sun while watching the stars. That was a beautiful night, sweetheart and I will never forget it. Gee it doesn't seem like seven years ago. Lonesome."

When rain came almost nightly, he slipped below deck into his bunk. The air was stifling, the sleeping accommodations, cramped. During the day, he savored idle moments in the music room writing to Flo, a fan a few feet away, drops of sweat punctuating the letter. When he had time, he reviewed his letters, crafted in neat handwriting, and carefully edited text using carets. One day his letter-writing was cut short following the unexpected call for a rehearsal on deck, when the military band worked on perfecting *Poet and Peasant Overture* by Franz von Suppé, an Austrian composer of the 19th century, and other pieces.

Guest musicians emerged from the crowd. They were members of three Army bands who asked to join the Coast Guard musicians in rehearsal. "They are for the most part very good," Art wrote.

An article in *The Salt* described the *Greely Grenadier*'s musical performance schedule. "They play marches in the morning, red-hot jam sessions in the afternoon or evening, and then pop concerts on Sundays, and the enthusiastic comments of the G.I.s and troop officers of crew members and gold braid alike, testify to the popularity of the music supplied daily by the *Greely Grenadiers*."[2]

Although music was the predominant entertainment on the *Greely*, Stedman planned a variety of events and activities to keep the crew and troops happy during long, monotonous days at sea. After the musical performances, movies were shown every night. *More Fun en Route for our Armed Forces*, a pamphlet provided to each passenger, illustrated dozens of games with playing cards, drawings, numbers, and words for solo or group players. Officers scheduled tournaments in checkers, poker, bridge, and chess. *The Salt* informed crew and passengers of events and tournaments, congratulated the winners, and profiled people of interest. Some crew members made up to $85 a game shooting craps, Art wrote, but he stuck to bridge and guarded his money. He sent Flo $35 of his monthly allotment, taking care not to squander the remaining $20. Large-scale gambling was not permitted on the ship, but gamblers got away with it. As the *Greely* steamed toward Calcutta on a later trip following the surrender of Japan, sailors were caught and reprimanded for gambling.

A few days after crossing the equator, Art got an additional assignment—movie and slide projector operator working under the education officer. The Eastman 16 mm projector was much like the one he had used while teaching music in Van Hornesville, except it had sound. He showed training and educational films and slides to crew and passengers, while another sailor worked the projector for the ship's evening movies.

As the heat beat down relentlessly on sailors and passengers, an event occurred as welcome as the sky opening and dumping cool rain on the *Greely*. The officers passed out one beer to each person aboard. Whatever Art might have thought about cantankerous Frank Boeberitz, he changed his tune and called him "a grand lad" after Boeberitz gave Art his beer. "I could have sold my bottle for $2 but wouldn't have parted with it for twice that amount—you know me, honey."

By the end of April, the temperature was moderating, but the seas were building. Art and George Bland grew mustaches, and Art noted the fuzz was growing back where he had been scalped. He spent a lazy Sunday playing chess with Roger Hartman and won. Recalling his Sunday walks with Flo, he wrote in a letter that would not be posted until he reached port, "What fun they were. Out here Sunday is just another day closer till I will see you again." By the end of the day, the seas were rough. Tom Stokes, chronically seasick, earned the nickname "Horizontal Tom" because of the position he assumed when seas rolled the ship.

Mascot Snafu seemed to take the seas well. She was featured in the ship's newsletter after suffering an accident. A cartoon depicted her in bed in sick bay. "She really is an awfully cute pup. Somebody stepped on her on deck, so they have a cast on her leg. Before we left port, the darned dog used to crawl into my sack every night."

In New York, Flo had grown accustomed to the near daily arrival of letters, with words of love, news of the band, and the wonders of the West Coast. They buoyed her mood in Art's absence. But after the *Greely* left port, the last few letters trickled in and then there was nothing. The mailbox that each day teased her emotions with anticipation was empty. She sank into a gloomy mood that she shared with the other *Greely* wives. Adding to the gloom was the continual news that the European war was about over, followed by news that it was not. "I'm giving in to the old ache again and I don't want to. Then they got us all pepped up with the V-Day rumor and keep saying it's only a matter of days and hours. Every time the radio fades, I expect to hear 'We interrupt this program to bring you a special bulletin etc.'" It was good news that Mussolini had been killed[3] but rumors that Hitler had met a similar fate turned out to be false. "Rumors are that Hitler is dead again," she wrote. "That man must have as many lives as a cat. I wish he could die that many times. It would be fun!"

After two weeks at sea with no letters, Art found himself talking to friends and strangers about Flo and wrote letters he knew would not be read for weeks. "I am actually getting very impatient to read your letters again dearest. It's bad enough being apart and then no mail either—that makes it worse." Loneliness is a condition that needs immediate remedies, but war rarely provides them. Flo was accustomed to receiving letters almost daily, the regularity conveying a sense of proximity. The musicians continued to write letters after the *Greely* put to sea but could not mail them for several weeks until the ship reached port. Intended to quell loneliness, their letters resided in envelopes oblivious of the urgency of their content. "Still no letters from you but I keep hoping," Flo wrote. "I guess you feel just as lonely as I do. I guess we're beginning to realize what the war means, eh?"

Before departing San Francisco, Art had sent her a photograph taken by the ship's photographer showing him, William Schallen, and three others, including Chaplain Hugh Miller singing while Harold Brody played the organ. Flo used the photograph as a bridge to span long, dark periods between letters. "Honey, I cannot tell you what effect that picture has on me when I'm home alone. I just run in and look at it every second. Finally, I took it into the kitchen so I could watch it. I must admit I did weep a little over it because you have one of your nice, natural expressions. You look so loving and swell I could just hug you."

The next day, the photograph caused mixed emotions as the reality of separation, the empty mailbox, and the inadequacy of the image sank in. "I don't know why I should have to be separated from such a cute mousie as you. Why do you have to be so nice? Why didn't you beat me and swear at me, so I'd be glad you were away? Why do you have to take such a cute picture, so I get more looking at you. It just isn't fair. You're very naughty, honey, that's all I can say. Very naughty." It may have been the photograph or the separation, but Flo belatedly and for the first time agreed to Art's plan to have three children after the war.

Greely Wives

Flo and Genny continued at odds, but they were getting used to living with minor conflicts. Flo sometimes sought the company of Francine, her best friend since childhood when the conflict heated up. Florence Cape and Francine Achenbach spent their early years together when the Capes lived in the Bronx. "We'll find something to do," Flo wrote of her seeking Francine and escaping the small apartment. "I have to do something to keep going on without you. If I go home, there's always Genny. We get along much better than we did but you know how it is… It also did my heart good that George [Bland] and Joe [Perna] think she's hard to get along with too." But Genny was a *Greely* wife, and the wives needed to stick together. After their first get-together at Lil Ketchel's, they made it a regular thing. They went to the Swiss Chalet for dinner, where their conversations skirted the edge of war. They laughed, enjoying fun and folly, and tried to forget the war. They topped off the evening at a bowling alley where the gutters claimed many of Flo's balls. She scored 60 and felt guilty about squandering three bucks. But the break was well worth it. "I always feel much better when I get out with that group," she wrote. "I felt so discouraged yesterday and they really gave me a lift. We all seem to feel closer to you fellows when we get together. We should do it more often." The next get-together would be at Flo's and Genny's apartment.

May 2, 1945. Dear Darling
They made an announcement tonite over the loud-speaker system that Hitler had just been shot in the streets of Berlin.[4] Guess everybody is mighty happy about that. I'd love to have been in N.Y.C. now to see the celebration. Our being way out here, the European war seems very remote, at least to me. If only you could see Geo. Bland and I with

moustaches, honey. As usual I expect an objection from you dearest, but I rather like it now. It's rather cool and we had quite a rough sea today so we didn't play on deck at all. Till noon today, we copied arrangements for the dance band and this afternoon, I helped run some educational movies for the crew. Frank's even acting quite sociable now; guess he is happy to be working with the electricians.

After the *Greely* docked in Melbourne May 4, the band members were given a map of the city and liberty. Four of them, including Art, went to town and got a steak dinner and some strong Australian beer. Back on the ship, he received seven letters from Flo dating back more than two weeks and two from his mother. His encrypted identification of the first port on the voyage provided Flo the information she needed to locate her missing husband, although the letter bearing the encrypted news and Art's reaction to Hitler's death would arrive a week later along with more than a dozen other letters written on the ship. Hitler died April 30, but it would be three days before newspaper headlines and radio bulletins screamed the news throughout the United States and on the *Greely*. Flo was not immediately aware of Hitler's death, despite the headlines. Or perhaps she regarded the news as another bit of false information she preferred not to hang her hopes on. Her May 2 letter, written at 10:12 p.m., does not refer to the biggest news of the war. In transcendent thoughts that span oceans, it was Art's ghostly presence, not Hitler's absence that dominated the message. "I do feel very close to you—and sometimes when I walk alone, I really think you're beside me. What a disappointment it is when I look up. Sometimes I make believe that you're there anyway."

A week later, Art responded to her fantasy. "Your letter was so adorable about you walking by yourself. Darling how I long to take you in my arms and tell you how much I love you." Flo would welcome the reunion with a condition. "Take a picture of yourself with that moustache and then kindly make with the Gillette. You know how I hate those things. If you come home with it, I'll refuse to recognize you."

While the *Greely* wives endured a long absence of letters, hoping that each day would bring letters with stories from the ship, the absence of Dear Darling letters was itself a message. Each day, Flo hoped that on her return home from work, she would find one but was disappointed. "I don't hope too hard though because I'm afraid there isn't. We decided that because we didn't get a letter by now that you didn't or wouldn't hit Hawaii."

She was right and the fears of the *Greely*'s wives were confirmed. A day before docking in Melbourne, Art learned of the *Greely*'s ultimate destination, at least three weeks away. The *Greely* would soon sail the most dangerous leg of the trip. The destination was encrypted in a letter.

May 3, 1945. Dear Darling
I can see right now that it is too early in the morning to write a good letter. I am seated on the floor of our music room trying to write with about ten guys milling around. The length of time I have to finish this is only about thirty minutes because we have a rehearsal. You can see why I am all screwed up now honey. As usual dearest when I am in hurry, I cannot think of a blame thing to say. I tried writing mom last night but it wasn't much of a letter, you had better drop her a few lines when you receive these letters. I think that I only wrote her four times during the trip but I couldn't think of too much that she'd be interested in. I am so anxious to get some mail from you and learn how everything is going, what you did about the car if anything.

CHAPTER THIRTEEN

Marriage and War

My parents and other *Greely* couples were intent on assuring spouses of fidelity during separation. My parents' letters are peppered with "love only you" and similar pledges. With that in mind, I began writing this chapter about marriage and war. The research revealed a dim picture of the institution of war marriage. But the phone call from my sister, Penny, in February 2023 brought home the damage that war inflicted on couples as no research could. Penny submitted a DNA sample for genetic testing several years ago, and in late 2022 received a communication from a woman in her 70s who also submitted a sample and obtained a report from Legacy Tree Ancestry indicating that she and Penny shared DNA. We had never heard of her. The report indicated that her mother was married to a serviceman who was away at war for an extended period and could not have been her biological father. During that time, she met one of our relatives, and in a brief relationship, conceived. When her mother's husband returned from war and found the household included a baby, he divorced her. She later remarried and the couple raised the girl. The biological father apparently did not acknowledge the daughter or the relationship.

Quite a few marriages during World War II failed under similar circumstances. It was a difficult time for marriages, with the divorce rate reaching an all-time high by the end of the war. Jane Mersky Leder, author of *Thanks for the Memories: Love, Sex, and World War II*, explains some of the reasons. "So many factors conspired to send couples to divorce courts: hasty marriages, long separations, battle fatigue, the newfound independence of women at work and in the home, children, and wartime passion mistaken as love."[1]

The forced separation of couples during the war fractured the traditional concept of marriage and ushered in a temporary and perhaps unwelcome sexual revolution.[2] Soldiers conditioned by the horrors of war returned physically and psychologically damaged, no longer the husbands who left home to join the military. Woman were lured from the household by propaganda seeking their assistance in war production. Posters of Rosie the Riveter welding and Rosie the Riveter flexing her bicep appealed to women's patriotic sense while employers offered income previously inaccessible to many women. No longer financially dependent on their husbands, working women earning a decent living and managing their lives found they could slip the bonds of marriage if they chose to.

The attack on Pearl Harbor prompted a rush to the altar. Some marriages were weak from the start, sparked by wartime emotions, the romance of men in uniform, and a draft exemption for married men. The United Service Organizations recruited women to do their patriotic duty by entertaining troops at USO dances and other social events, where a young woman could meet hundreds of men and sometimes an instant husband. The USO instructed female volunteers to be sweet, clean, and cheerful, as troops preferred women who looked like girls.[3] After promoting encounters between young women and troops, the USO put the brakes on wartime romance. *Hail Hostess*, the USO's bulletin, cautioned, "Be realistic about romance! Romance is wonderful in its place, but the USO is hardly ever the perfect place. For one thing, the boys have a lot of things on their minds, but you are probably not one of them. They love fun but they are frightened to death of entangling alliances. And incidentally, an amazing lot of them have dear little wives and sweethearts back home. (They may neglect to mention this but take our word for it!)"[4]

Before the United States entered the war, the Selective Service and Training Bill of September 16, 1940, required all men between the ages of 21 and 35 years to register for the draft. Congress amended the Selective Service Act to allow married men a deferment from Class I-A draft classification until further notice.[5] There were exceptions. Men who were aware of their impending induction when they got married were not eligible for deferment under the new amendment. Married men who had been inducted into the armed forces were ineligible for a deferment. *Life magazine* reported that only 18,700 men had registered for the draft by the end of 1940. A far greater number headed for the altar and a possible marriage deferment. The marriage rate for females over 14 years of age jumped from 69.9 per thousand in 1938 to 93 per thousand in 1942. In 1943, *Keystone*, the jewelry trade magazine, reported that the supply of wedding rings was dangerously low because of the rush to marry.[6] The motive and the haste did not bode well for many marriages and may explain some of the divorces by the war's end.

In only a brief span of time, war shattered norms of sexual behavior. Incidences of infidelity, which increased significantly during the war, created rifts in traditional families, justified divorces, increased the risk of venereal diseases, and invited social condemnation. In the culture of wartime infidelity, women suffered the brunt of condemnation, the difference in sexual standards reflecting the longstanding double standards that men's infidelity was acceptable while women's was not. This difference was enhanced during the war because men were required to serve in distant and dangerous lands, risking their lives while women remained in the relative security of the country. The double standard manifested itself on military bases, where condoms were available for male soldiers while the Women's Army Corps urged abstinence. Propaganda posters of the time blamed women for spreading venereal diseases. One poster showed a young woman with a girl-next-door face. Just under her face, the caption read, "SHE MAY LOOK CLEAN – BUT," followed by "Pick-ups, 'Good time' girls, Prostitutes spread Syphilis and Gonorrhea. You can't beat the Axis if you get VD." News magazines of every type published worrisome articles about an epidemic of venereal diseases, especially among single women. According to the *Ladies' Home Journal* in 1944, about 11,000 girls between the ages of 11 and 15 were infected with syphilis. The U.S. Army's studies found as much reason to blame men as women for infidelity and the transmission of venereal diseases. A survey conducted in 1945 found that 80 percent of soldiers who served away from home for two years or more reported regular sexual intercourse. Almost a third of them were married.[7] When the war ended, reunions began, but recollections of servicemen or wives of their brief, prewar romances were, in many cases, fantasies. Author John McPartland wrote, "The war ended. Combat men began to go home, unaware of the deeper currents that lay beneath their dalliance in Europe. They returned to their one-week brides of training camp marriages, their sweethearts, their wives and children."[8]

Some judges overseeing divorce cases during the war attempted to crack down on infidelity, largely blaming women. Cook County Illinois Prosecutor William J. Tuohy announced that he would prosecute wayward wives or husbands who were parties in divorce cases involving infidelity. His actions were prompted by the divorce of war hero Cpl. Stanley Heck, 30, and wife, Henrietta Heck, 26. High school sweethearts, the couple married in 1939. In 1943, Heck joined the Army, served in the invasion of Germany, lost his legs, and mangled his hand in a landmine explosion. He lost his wife to meat buyer Alvin Schupp, a married man who employed her.[9]

Among the many additional cases Tuohy cited was that of Yeoman 3/c Roy Popularum. After serving in the Pacific, Popularum returned to his home in Chicago to find his wife, Dorothy, and their son living with a man who was wearing Popularum's clothes. The stories are legion, the remedies, few and controversial. Tuohy earned praise for punishing adulterers with sentences ranging from a $500 fine or a year in prison or both for first-time offenders. Those active enough to reach the third offense faced a $3,000 fine or three years in prison or both. In Newark, New Jersey, Judge P. James Pelleccia Jr. had come across twenty cases of wives' infidelity in military marriages. His wistful remedy—"If I had my way, soldiers' wives who are unfaithful would be branded with the scarlet letter and had their heads shaven."[10]

The number of failed marriages by the end of the war illustrates the problem of separation and other factors generated by war. By 1945, the divorce rate was 3.5, or 3.5 divorces per 1,000 marriages, an all-time high up to that year and almost twice the divorce rate of 2.0 in 1940. In 1946, the divorce rate jumped to 4.3 before dropping and not reaching that peak again until 1976, when it climbed up to 5.[11]

The change in sexual standards was evident in sexual conduct outside the institution of marriage. Incidents of sexual assault charges were 27 percent higher in 1944 than in prewar averages.[12] The sexual revolution among single people also was evident in an increase in pregnancies out of wedlock. The ratio of out-of-wedlock births to married births doubled in just two years, with one in twelve recorded in 1944. The Child Welfare League reported an increasing number of women keeping their babies rather than arranging adoptions. Giving up babies for adoption raised complex problems for married women whose husbands were at war for extended terms. Thorny legal questions arose when the rights of absentee husbands became entangled with the rights of biological fathers and remained unresolved during the war.[13]

Not Rosie the Riveter

Florence Schnell was not a Rosie the Riveter recruit. She did not go to work because of the war. Florence Cape spent her early years with her family in the Bronx near Arthur Avenue, then an Italian neighborhood, now an upscale commercial district of delis and boutiques. She was the oldest of three. When she was 11, the family moved to Valhalla, a hamlet north of Manhattan where she enrolled in 6th grade and pursued her interest in music. She played trombone. Music was not her best subject; her music grades were mostly Bs and B+s. She excelled in history and in the eighth grade took first place in a *New York Times* essay-writing competition about the personalities of the constitutional convention. When she prepared to enroll in high school, her musical aspirations hit a snag. Girls were not allowed in the high school music program at Valhalla High School. Minnie Cape, adamant that her daughter pursue a musical career, aggressively petitioned the White Plains High School to accept Florence as the high school music program accepted girls. Flo graduated in 1935 with a high school diploma in music.

She acquired a love of music at a young age from an unlikely source—her father. Many studies have explored the characteristics of musicians, finding them agreeable, conscientious, sensitive, and open to new experiences.[14] If that paradigm is valid, her father, George Cape, was an anomaly. He was a rough, mean man by all accounts including my memories.

When I knew George, he and Minnie lived in Youngsville in the southern Catskills. He drove daily to New York City to work in construction. When he wasn't working or driving, he was cussing, drinking, denigrating people who were not like him, denigrating people who were like him, fishing, killing animals, shooting at something. He was devoid of empathy for most life forms including humans. The farmhouse in the Catskills was a menagerie of trophy killings. Deer heads hung on the walls in every room. Antlers were stored in most dresser drawers. A one-inch cork plugged a hole in the dining room chair rail from an accidental 12-gauge shotgun discharge. Empty beer cans, some with bullet holes, decorated the landscape. In the living room, the giant head of a black bear, its mouth agape, seemed ready to perform an aria when the piano beneath it came to life.

And it did. Each time we visited the farmhouse in the Catskills, my mother would gather us in the living room, seat us in chairs arranged as in a miniature concert hall, and coax George to play the piano. He'd sputter and fain some excuse until we could wait no longer. Then, like a modern rock star holding a venue of fans hostage, he'd enter the living room, take his seat at the piano, and play, oblivious to his audience. The music was largely popular tunes of the thirties and forties, show tunes, or big-band music adapted to piano. When he tried to end a performance, my mother would coax him to play one more tune, then another, and another. The music sounded pretty good to me. Finally, he would get up while my mother whipped up an applause, and without a hint of acknowledgment, he would walk out, crack open a beer or guzzle a bit of whiskey. It is unclear whether my mother arranged these performances to instill in us a love of music or to demonstrate that her father had some redeeming qualities. Or both. Years later, in the attic in the New York house, I found a box of musical scores of Broadway shows and other popular pieces dating back to World War I. Still convinced that George did not have an ounce of refinement in him, I asked my brother where they came from. He was certain they had belonged to George. Later I found his name handwritten on one of the scores. Still hard to believe.

Flo's father had little use for colleges or people who attended them. Fred, her brother, and Adele, her sister, did not attend college. Adele married young and spent most of her life as a housewife. Fred joined the Marines and served in Guadalcanal, where he was injured. With the generosity of an aunt, Flo enrolled in music education at Ithaca College, where she met Arthur Schnell, also a music major and trombone player. They bonded over music and married in a ceremony near the Catskills farmhouse June 29, 1940, more than a year before the United States entered the war.

Flo and Art recognized the risks of sexual misconduct and other impediments to marriage and took measures to assure each other there would be no breach of marital vows during separation, war, and in the future, using both promises and anecdotes conveyed through the only means they had during much of the war—letters. During the Manhattan Beach era, the couple were separated for only a few months. Nevertheless, assurances flowed almost daily. By expressing his disdain for the reaction of other sailors when thirty SPARS from Hunter College passed his Manhattan Beach barrack, Art quelled any thoughts Flo might have that he had a roving eye. On the train, after relating the story of the Chicago woman who hopped in the sack with four flyboys, he assured her he would not do such a thing.

Once Art was transferred to San Francisco, the separation became dire, and assurances, more frequent, the reasons understandable. He and his buddies were out on the town many nights, frequenting restaurants, music clubs, stores, having a jolly time. He estimated the ratio of men to women was a comforting (for his wife) ten to one and concluded most of his letters with reassuring phrases such as "I am lonely only for you" and pledges of exclusive love. Many of Flo's letters are similarly closed. Art was pleased that the abundance of letters he received was considered an indication of his wife's loyalty.

Officers of the U.S.S. *General A. W. Greely* were undeterred by the shortage of women, their big ship a floating venue for sexual fraternization. The last day of March 1945, they assembled 150 SPARS and WAVES (women enlisted in the Coast Guard and the Navy, respectively) and boarded the *Greely* for a sight-seeing cruise around San Francisco Bay. Art maintained he kept a distance from the passengers. "Don't worry, Mousie. They won't bother me."

Art was a handsome fellow, with wavy black hair that shined like polished boots. Soft, blue eyes the color of the sky at twilight returned a warm gaze. In his letters he projected a gentle personality, avoiding criticism, harsh language, obscenities, and negative comments except when discussing musicians he did not like. Playing love-struck melodies in the dance band and rousing patriotic tunes in the big band to soldiers and sailors, including fifty Red Cross women, presumably single, would be enough to make any war wife wary. Perhaps he understood that when he wrote to Flo on April 12 that the *Greely* would carry fifty Red Cross men. Is it possible he did not know they were women? Commander Stedman referred to Red Cross women at the time of the commissioning. More likely, Art wanted to reassure Flo there were no threats to their marriage as he was aboard a ship full of men. The cover lasted until he sent her copies of a cartoon by George Bland, depicting women on the *Greely*. He came clean when writing about the Ancient Order of the Deep ceremony, when he described the Red Cross women taking part in the ceremony wearing bathing suits.

The *Greely Grenadiers* were the big attraction each day and evening, surrounded by fans. The Salt reported in an edition Art sent to Flo,

> Wherever and whenever they play, and whether it's the smaller "jive quintet" or the full twenty-six-piece band, the *Greely* musicians were hemmed in like an American girl by wolves at our outbound destination. In fact, sometimes the bandsmen are lucky to escape with their instruments after a show. It is seldom indeed that the crowd, which almost always includes Captain Stedman, does not demand additional numbers.[15]

Flo's letters conveyed pledges of fidelity similar to Art's. "Be a good boy, Honey, and I'll be thinking of you every single second. Bye for now, sweetheart. Love only you forever," she wrote the day he boarded the train for San Francisco. She also used anecdotes to inform him about her views on wartime infidelity. Her neighbor, Mrs. Herman, had a friend whose husband was fighting in Italy. Flo learned the friend was going out with another man. "I would like to tell her a thing or two," she wrote. "The heel!"

While men of military age on the home front were greatly outnumbered by women with limited access to sexual partners, sailors seeking female companionship were sure to find it at most ports in the Pacific and Indian oceans. Calcutta and other ports were a sexual free-for-all for servicemen willing to take the risk of venereal disease. Years after the war, my father told me that Indian kids would dive deep for pennies, and single men could have their pick of young women for about the same price. The American Red Cross *Guidebook to Calcutta, Agra, Karachi and Bombay* provided to sailors and passengers of the *Greely*, explained the risks of venereal disease. "The best preventive is abstaining," the guidebook warned with little confidence as it continued, "There is no sure method of prophylaxis. Nevertheless, prophylactic measures markedly reduce venereal incidence. All persons exposed are, therefore, urgently advised to use prophylactic treatment immediately following exposure."[16] In late August, a dozen troops were caught frequenting a brothel outside the approved liberty zone and were confined to the ship. No musicians were among them.[17] Elsewhere, troops were a big hit with women in foreign countries. "These foreign women must be forgiven their surrenders," John McPartland wrote in Coronet magazine. "Swaggering hundreds of thousands of men, confident, well fed, comparatively rich, and avid lovemakers swept through their towns. The Americans believed in direct, brutal sex combined with the sort of romance invented by the movies; women responded to that kind of combination."[18]

Rumors that the *Greely* musicians' sea duty would be brief were comforting to the *Greely* couples. But they were wrong. Clare Grundman's position that the musicians would serve one voyage, enough to get a campaign bar and then go home, was wrong. *Greely* Chief Musician William Schallen's rumor that with the addition of twelve musicians, the *Greely* Grenadiers could expect shore duty was wrong. Only three additional musicians joined the original band. Some rumors of the *Greely*'s destinations were wrong. The ship did not initially sail between Europe and New York to bring home troops from the European war. She did not sail for Hawaii. But after more than half a year at sea, Art returned home to Flo, George Bland to wife, Betty, Belmont Ketchel to Lil, Frank Boeberitz to Genny, Joe Perna to May, Danny Cowan to Natalie, Jack Purves to Chris, Shelly Manne to Flip, Arnold Broido to Lucille, Harold Sachs to Miriam, and hundreds of other *Greely* crew to their wives, their marriages having survived the most catastrophic assault on the institution up to that time. They all were changed by war. But they were fortunate that the duration of separation was relatively brief, their battle scars, etched by anxiety rather than weapons, their wives, carrying on in their absence, stronger, more independent than when their husbands left but waiting, nevertheless. Before those reunions would take place, the *Greely* crew and passengers had a couple more voyages to make through hostile waters and turbulent seas, performing time-honored nautical rites of passage while attending musical performances, playing bridge, poker, checkers, shooting craps, gambling, grumbling, dancing knowing that in an instant their good times could turn deadly. Although they did not mention it in correspondence, the *Greely* wives knew it too.

Alex Haley, author of *The Autobiography of Malcolm X* and Pulitzer Prize-winning novel *Roots*, was the Coast Guard's chief journalist. He received the Coast Guard Academy's first honorary degree and was the namesake of a Coast Guard cutter.

CHAPTER FOURTEEN

Australia

From Melbourne, the U.S.S. *General A.W. Greely* sailed into bad weather. A day out, she was in the middle of a major storm, taking seas over the bow, pitching and rolling like a cork in a washing machine. The music room resembled a makeshift hospital, with men lying around in every available space looking like they were dead or dying. Despite the large seas and high winds, Art did not get seasick, although after a while he felt groggy and tired. The heat became unbearable. He didn't feel much like writing letters, but he managed to pen one to Flo and one to his mother describing the storm, which he seemed to endure better than he endured the heat. The musicians spotted land and knew docking was imminent, but the name of the port was not disclosed. Art thought he knew where they were.

May 8, 1945. Dear Darling
I promised myself to write a lot of people letters but you, darling, are the only one I really enjoy writing to. We expect to make a more or less unscheduled stop at a port soon, so the letters I have written since the last port will get mailed. There really isn't too much news to write about this time except this morning we heard that Germany surrendered unconditionally. What times the people will be having for themselves in N.Y. Gee Honey how I wish we were together to celebrate that.

The *Greely* War Diary and deck log do not list Perth as a port of call. This stop was Freemantle, where the *Greely* docked May 10. For the censors, Art called it "Some port in the Pacific." The Perth mistake gave her a location to cling to, the discrepancy, irrelevant by the time the letter reached her. Freemantle was a one-day stop, no liberty for the crew or passengers.

"It seems like we just stop long enough to get a teasing glimpse of land and away we go... It was not without its pleasant surprises, though, honey, because we unexpectedly got some mail aboard, which included some for me." He received five letters from Flo and put some in the military mail for her. None of the letters convey V-E Day exuberance. The war was not over for him, the musicians, or the troops on the ship, and the *Greely* was heading for conflict. The risks were well known. Just a month earlier, the Allies suffered 50,000 casualties as they clawed out a staging ground on Okinawa for Operation Downfall, the invasion of Japan. President Truman noted in his V-E Day speech, "But in other homes, while there will be joy and anticipation, it will be dampened by the thoughts that their loved ones are still waging a bitter battle in the Pacific area."[1] For many of the crew and troops on the *Greely*, the war was just beginning.

A day out from Freemantle, the dance band was rehearsing on deck when a rogue wave rolled the ship. "Some of the fellows fell off their chairs; others came sliding across the room right into the wall—luckily no instruments were damaged but that sure broke the rehearsal." Later, a few friends crashed Art's letter writing session and sent their regards directly to Flo in his letter. "How I wish the four of us were sitting around a card table again!" George Bland wrote. He, Art, and horizontal Tom Stokes had enjoyed lots of laughs, he told Flo. "You can bet we miss our ever-lovin' wives like crazy." Tom Stokes expressed similar thoughts and promised to be home soon. Art closed with a warning based on the Blands' expecting a baby. "Take a tip from what happened to Betty Bland and beware when I get home."

A few days before the *Greely* arrived in Freemantle, the *Greely* wives were on edge, expecting an announcement any day that Germany had surrendered, news they would love to have shared with their husbands. "Remember Honey we were going to celebrate V-E Day together. Little did we know," she wrote a day before the celebration. "You were going to get me plastered, tsk tsk. Imagine me plastered. I'll probably weep a quart though—I feel like it right now. How would you like to have one of those weepy women? I promise I won't cry if you'll only come home. Honest, honey." Her emotions soared and tumbled with each bit of good news, with each false rumor of peace. She reacted to one V-E Day false rumor after another. Some parents misinterpreted early news reports of victory in Europe and came to the school to pick up their kids to celebrate. The principle turned on the news in time to hear "unconditional surrender." "Well the kids were jumping up & down and we were all half weeping. And then about ten minutes later it was Italy. What a letdown."

Flo and her coworkers weren't motivated to teach classes after the news of Hitler's death, given the excitement of Germany's surrender. But they were required to work. She and three other teachers snatched a few moments of hooky, snuck out to a car, and listened electrified in anticipation, in communion with millions around the world, waiting for the radio to spill the news. Still no word. They drove around for ice cream sundaes and brought one back for the principal so she would not be peeved at them. Back in school, they occasionally left their classes to check the car radio and returned to work. They knew the announcement was coming, along with a speech by President Harry Truman.

Speaking from the radio room at the White House, President Truman announced Germany's unconditional surrender. V-E Day (Victory in Europe) was celebrated around the world on May 8, 1945. As Flo and Genny were not on speaking terms, Flo planned to celebrate V-E Day with her best friend, Fran, suggesting they go to Radio City. But after talking by phone the day before with Fran's mother, Francine, she had second thoughts. "She kept saying, 'don't celebrate, don't go out—think of the blood that has been shed.' She kept it up until I thought I'd scream." Flo did not explain Francine Achenbach's intensely negative reaction to celebrating V-E Day, surprising considering her son-in-law, George, was fighting for the Allies in Europe. The Achenbachs were of German ancestry and close to the German community in the Bronx. They ran a German bakery. Among their patrons was a German immigrant named Bruno Richard Hauptmann. In 1932, Hauptmann was arrested and charged with kidnapping and killing the 20-month-old baby of Charles Lindbergh Jr. and Anne Morrow Lindbergh. After a five-week trial, he was convicted of first-degree murder and later executed. Among the witnesses were the Achenbachs, who believed testifying that Hauptmann had entered the bakery feeling ill days before the crime would help exonerate him. But prosecutors used their testimony to help win the case. Many patrons, believing Hauptmann innocent, blamed the Achenbachs for aiding in the conviction. The bakery business withered. Francine may have kept a low profile during the V-E Day celebrations to avoid further ostracization of the German community. Maybe Francine didn't like confetti or crowds. Or maybe she didn't want to celebrate the homeland of her ancestors having been blown to rubble. Perhaps it was, as she said, the enormous blood shed by Allies and enemies. Whatever her reason, it is important to realize that not everyone in the United States openly celebrated the end of war in Europe.

Flo ignored Francine's advice and, together with Fran and her cousin, headed for Time Square. She described the celebration. "Boy Honey it was fun!! The place was jammed, and people were throwing confetti around & blowing horns. They had all the lights on. The brownout was taken off and no traffic in the streets. We swarmed all over." Fran's husband had been away at war two years to the day, and Flo believed that if he didn't come home soon, Fran would go berserk. After the celebration, Flo joined Fran and her cousin window shopping for black-lace nightgowns to prepare for their husbands' homecoming. Over in Central Park, the remnants of the Manhattan Beach Coast Guard Band performed for V-E Day. Flo didn't go but learned from a friend the band sounded pretty good. She told Art that "Chief Mulder got up for a bow. Grrrr." A month later, Art reacted to the news about Mulder with resentment. "So Mulder's outfit played for V-E Day ceremonies—that's almost as bad as having a German band play—who me, bitter?" With no timely letters from Art, Flo tried to imagine where the ship was during the V-E Day celebrations. "I try to visualize where you are & try to picture nothing but water & I just get sick."

At home, the wartime shortages of food, gasoline, and other commodities were taking a toll on the *Greely* wives. While the federal government put money into people's pockets, its vast consumption of products and materials once available to consumers created scarcity even for those with money. Sugar, coffee, meat, and other staples were hard to come by. Meat was so scarce that some restaurants began offering buffalo or antelope steaks, or even beaver burgers. Flo complained to Art about the crunch in the meat supply. Cars were scarce too. Detroit was not making cars. With the conversion to war economy, automakers had retooled to produce about a fifth of all war materials. Henry Ford produced 1,200 B-24 bombers in a mile-long plant covering sixty-seven acres instead of building cars. The supply of rubber for tires dwindled after Japan seized the Dutch East Indies and Malaya, cutting off 90 percent of American rubber. Flo considered selling their car, but her mechanic advised her to keep it. The shortage of automobiles would continue after the war, he warned. She kept it.

Flo was busy teaching at Franklin Square Public Schools and working a second job, sometimes skipping one for the other, and rehearsing for a glee club performance. She also directed music for a school Christmas play. As the school year wound down, her musical career gained momentum. She was invited to serve as chairwoman of a musical contest for public school musicians held in Farmington, Long Island, where she met musicians she and Art had known from Ithaca College. Many of them asked about Art, who was thrilled to hear of his wife's success. "Honey, I need you more than anything else to be a success in this world," he assured her. "Whatever success I have had so far has been due directly to you, and it will be that way in the future too."

Relations with Genny continued to sour. The apartment felt like an oven, with Genny claiming the draft of a fan made her ill. She said she would leave the apartment and live with her parents if Flo did not travel to the West Coast when the *Greely* returned from its first voyage. That was good news to Art. He worried that if he and Frank were discharged at the same time, the two couples might end up sharing the apartment. The simmering conflict would surely erupt into fireworks. Flo and Genny got along well enough to take a short car ride to the laundry service, passing by the Manhattan Beach Coast Guard Training Station. The remnant of the band was playing colors as they drove by. "Honestly I could have wept," Flo wrote. "They didn't sound as well as your outfit although they really weren't too bad, but they weren't [marching] too straight on the corners." She imagined Art as part of the band, coming through the gate, meeting up as they had done only six months earlier.

As the *Greely* entered the Indian Ocean on May 14 the danger of Japanese submarine attacks mounted. The ship passed within 90 miles of the Coco Islands, a group of coral atolls covering just over fourteen square miles, some 650 miles south of Sumatra (now Indonesia). The island, controlled at the time by the British, was attacked by the Germans in World War I and by the Japanese in World War II. In May 1942, Sri Lankan soldiers patrolling the island for the British attempted a mutiny. They planned to turn the island over to Japanese forces that had occupied the nearby Christmas Islands a few months earlier. Of the seven convicted in the failed mutiny, three were hanged. Although Japanese submarine operations were considerably weakened by the time the *Greely* sailed the Pacific and Indian oceans, the subs continued to be a threat. A destroyer sailing with the *Greely* off the coast of Sumatra located an enemy sub a quarter mile ahead and dropped depth charges. On May 16, the British destroyer H.M.S. *Relentless* sailed north from the Equator to Colombo, Ceylon (now Sri Lanka), to join convoy OW5/2 traveling with the *Greely* to provide security.

Despite the dangers in the Indian Ocean, *Greely* officers conducted another Ancient Order of the Deep ceremony for those who had not yet crossed the equator. With the destination imminent, Art wrote an encrypted letter.

May 17, 1945. Dear Darling
I can't imagine how it would be to see a little snow now. From all reports, New England had a very bad snowstorm. I'm looking forward to some liberty in this next port, as I want to do a little souvenir hunting. They can't charge high prices though or I won't be getting much because my pay is pretty small. Something unusual happened last nite, I got real ambitious and wrote a letter to Mom, Bill and Fran. It's time I received a letter from Bill as I haven't heard a word from him since we left the States. I'd try if I were you dear to go to Ocean Grove with Fran for a week, it will do you good to get away for a while. Sweetheart, all your letters are wonderful but the one of April 26 was perfect.

The *Greely* reached the mouth of the Hooghly River, a broad, navigable river flowing south from destination Calcutta. At 2 a.m. the crew dropped anchor at the mouth of the river to board a local pilot. As the *Greely* neared its destination, Commander Stedman addressed the passengers.

Tonight is our last night at sea. God willing and fair weather we will arrive at our destination tomorrow. In just a few minutes you will hear *Taps*—the last time you will hear them on the *Greely*.

As commanding officer and on behalf of the officers and men, I wish to say it has been an honor, a privilege, and a pleasure to have been assigned the task of bringing you from the United States to this theater of war.

We on the *Greely* have been happy to have you aboard. We have tried in our small way to do everything we could to help you enjoy this trip and make it a memorable one. And now, as always, all good things must come to an end and the best of friends must part.

You all have a long hard road ahead of you with death, disease, and many heartaches staring you in the face. You will come to the sudden realization that this is a war and a very cruel war that gives no quarter and asks no quarter, but when you have completed your tour of duty—and believe me it will be a tough one—one which you and I and the whole world will be proud of—we here on the *Greely* will be waiting with open arms to receive you aboard and take you back to your loved ones.

A brass quartet performed "The End of a Perfect Day" followed by *Taps*, the music piped throughout the ship.[2]

Arriving at a secure destination was a relief for musicians, but bad news quickly followed—no mail had arrived. Art and his buddies were granted liberty and took a tour with the Red Cross. It was not a pretty sightseeing tour. Sacred cows roamed the streets and crows as thick as pigeons in downtown Rochester soared through the filthy air. A cholera epidemic claimed thousands of lives a day. The site where bodies were burned did not seem like a tourist attraction, but it was on the Red Cross tour. The heat was unbearable. Sailors drank only heavily chlorinated water at the Red Cross facility. Amid the devastation, the musicians saw palatial buildings.

Back at the ship, 2,923 troops and civilians debarked. When the musicians returned from the tour, more than 3,000 troops were ready to board and go home. Some had served more than three years in the Pacific Theater. *The Salt* newsletter featured two soldiers who had lived among head-hunting Nagas, a primitive tribe of warriors in the Naga hills in northeastern India. The pair were members of a Signal Corps team that established signal aircraft warning stations in the wilds of northeast India. After a week-long hike, the soldiers arrived at a location to construct a station. Although the Naga spoke no English, the soldiers learned to barter with them to help build the signal units. During their stay in the jungle, the Naga raided a village and returned with their trophies, the gruesome details of which were suggested but omitted from the newsletter. The soldiers left the location with photographs and artifacts including weapons used to dismember bodies. The weapons were illustrated in the newsletter by cartoonist George Bland.

No food supplies were loaded in Calcutta as filth and disease were ubiquitous. Armed guards transferred four soldiers from the Calcutta-Burma stockade to the *Greely*, where they were confined to the brig during the passage.[3]

Sailing south at 15 knots from the mouth of the Hooghly River, the *Greely* was back in dangerous waters. Radar picked up numerous ships, all friendly, and escort aircraft that scanned the waters for submarines. On May 29, radar detected three friendly aircraft that soon faded from radar, replaced by another that briefly followed the ship. My father described the relief at seeing British Seafires (the naval version of the Spitfire) emerging from the horizon, at times flying overhead. "It sure made us feel good to see those guys," he told me years later. The sailors cheered and waved their hats to the pilots, who dipped their wings in recognition. But minutes later, the drone of the engine subsided. The fading image pierced the horizon. Fear and anxiety quietly settled back into the collective psyche of sailors. Art advised Flo of the destination.

May 26, 1945. Dear Darling
(The code cue appeared in the middle of the first page.)
I can see your face now cutie wondering why I am making two letters out of one. We even had movies aboard tonight, a very corny show about the Northwest Mounted Police. Can't you just see me enjoying

that, sitting packed in like sardines. Lonesome, lonesome, honey—it has been almost three months we've been apart. If only we have a little time off when we reach the States so that we can go places and do things. I never realized I could miss you as much as I do now, darling, I love you more each day.

The *Greely* docked at Ceylon to load food and other provisions for the return trip and enjoy a respite from the threat of a torpedo attack. As the ship approached land, the musicians observed a dramatic mountainous terrain and, as the ship got closer, a clean city. The *Greely* loaded provisions at Ceylon, which was well stocked with food and other supplies. No one got liberty. After departing, the *Greely* served some of the finest meals at sea followed by cherry pie and ice cream. Art drank his first Pepsi Cola since leaving California. There was so much he wanted to tell Flo about India and other ports they visited, but the censors were standing by to slice the paper or confiscate the entire letter.

In the summer of 1939, the U.S. Lighthouse
Service was consolidated into the Coast Guard,
resulting in the transfer of 5,200 personnel,
as well as ships, depots, and district offices from
the Commerce Department. The merger meant
that the Coast Guard was responsible for
maintaining 30,000 aids to navigation—buoys, day
markers, radio beacons, lightships, and lighthouses
along America's seacoasts, lakeshores, and river
systems in peacetime and war.

CHAPTER FIFTEEN

Censorship

In the early days of the war, a homesick merchant sailor, in a risky North Atlantic crossing, devised a way to communicate his port destination to his wife.

"Dearest," he wrote from the ship, "our convoy is three days out of the United States. As you know, we are not supposed to tell where we are going, but I happened to think that we could work up a little code of our own so that I could let you know without anyone else being the wiser. If I mention in a letter that I saw Mabel, you will know that we docked at Liverpool. If I say I saw Ruth, you will know that we are in Glasgow. Catherine will mean Iceland, and Helen—Heaven help us!—will mean Murmansk." Confident that his code was in place, the sailor later penned a letter from Glasgow. "I had a date with Ruth last night, and I expect to see her for three or four more days."[1]

The next month, the mail from the States arrived along with a letter from the merchant seaman's wife, who considered Ruth to be his paramour. "Of all the nerve," she wrote. "Who is Ruth, and what do you mean by going out with her? Have you forgotten that you are a married man? Or don't you love me anymore?"

Neither husband nor wife knew that censors had intercepted the code letter, leaving the merchant seaman with the unenviable task of explaining to his wife what he really meant by having a date with Ruth. The code they used, similar to the code Art, Frank Boeberitz, and other sailors devised, was known as "open code," a system understood by writer and reader to provide seemingly innocent and harmless information to a spouse or other loved one. Theodore F. Koop, deputy director of the Office of Censorship and

Seal of the Navy censor indicates the contents have been reviewed and redacted if necessary.

author of *Weapon of Silence*, wrote that detecting open code was among the most challenging of the censors' responsibilities.

"Open codes usually stood out because they resulted in stilted language of unusual context. Most of those discovered were used, not by spies but by good Americans who thought they were 'beating the censor' because of a law that no unapproved codes could be used in international communications. They thought they could quietly pass along a little military information to each other, and who would be the wiser? Nobody, in their opinion, unless the recipient talked too much—but it was always the other fellow who did that."[2]

In scrutinizing international cables for open code, censors had the upper hand. With only suspicion of open code, they paraphrased the message, thereby scrambling the code, a strategy that would not work for letters.[3] Seizure of suspicious letters was the only option and in most cases, a dubious strategy. Censors became aware that cable orders for flowers could be open code with sailors and soldiers informing their wives or sweethearts of their location by the type of flower. Censors struck out the flower type, letting the florist make the choice. The delivery date was left open. This brand of open code was believed to be one of the oldest and most popular.[4]

Like the merchant sailor, Art sent his first code letter through the ship's mail March 14. Concerned that it was intercepted by the ship's censors, as was the case with the merchant seaman, he mailed a second code letter the same day through the Post Office in San Francisco, which was censoring only international mail. Both made it to Brooklyn. Other *Greely* musicians did the same, helping each other beat the censors by carrying mail to the Post Office when on liberty. Art asked Flo to confirm receipt of the code letter, which she did. "I didn't understand what you meant in your Friday letter—Dear Darling—but have it all straight now." The communication system was set up and the censors were deceived. As far as I can tell none of Art's open-code letters was censored. Art and other sailors considered the practice of revealing their location by code harmless, but in enemy hands information about the location and movement of ships and troops, even in the recent past, was considered dangerous as it could indicate military strategies, troop buildup, and sailing routes. "Many men in the armed services or in the merchant marine were as thoughtless as the sailor who prided himself on devising his code of girls' names. Suppose his wife, at the bridge table, had mentioned his convoy's arrival in Glasgow and had been able to add details of its further route. Was the enemy listening? There was a court-martial ahead for the soldier, sailor, or marine who passed along military information by the open-code method."[5]

Censors scanned mail and cables for any suspicious reference to military or war. They removed references to the Bible, "xxx" for kisses, college students' grades, suspicious numbers, and poems. They found messages in postage stamps included in envelopes for return mail and messages under postage stamps. They even intercepted letters with invisible ink that revealed ship movement once they were treated with a chemical.

Before the United States entered the war, before the Office of Censorship was formed, censorship of correspondence was practiced. FBI Director J. Edgar Hoover authorized his agents to work with the Post Office to monitor the mail of resident aliens the FBI suspected of having contact with hostile countries. As early as the fall of 1940, FBI agents were surveilling the correspondence of immigrant Japanese leaders, although the agency had no statutory authority to do so, and the immigrants likely had no access to sensitive information. The United States was also on the lookout for messages suggesting or detailing injustices in detention camps forbidden by the Geneva Convention. "Should such information slip into the wrong hands, it might lead to repercussions against interned citizens of allied countries held in enemy territories. Camp administrators were continually vigilant about conforming with the terms of this international agreement regarding all aspects of captivity."[6]

In response to growing hostilities in Europe in mid-1930, the military

Florence Schnell expresses her embarrassment knowing that a censor has read her husband's risqué letter.

began developing a censorship program under the authority of an Army-Navy board. The Military Intelligence Service would oversee international mail and domestic telegraph lines while Naval intelligence would censor radio, international cable, and overseas telephone lines under the direction of Commander H.K. Fenn. By October 1941, Fenn had trained 346 reserve officers who then returned to their civilian jobs, awaiting orders to begin military service. With the attack on Pearl Harbor, the sleeper censorship program awoke and became an active initiative. A week later, Secretary of War Henry Stimson gave the order to begin censoring operations within the Post Office.[7]

President Roosevelt had a more comprehensive censorship program in mind when he enacted the Office of Censorship in December 1941. He appointed Byron Price director and gave him sweeping authority. Roosevelt's Executive Order read, "The Director of Censorship shall cause to be censored, in his absolute discretion, communications by mail, cable, radio, or other means of transmission passing between the United States and any foreign country or which may be carried by any vessel or other means of transportation touching at any port, place, or Territory of the United States and bound to or from any foreign country."[8]

Roosevelt knew his appointee was in for a thankless job. "All Americans abhor censorship, just as they abhor war," the president noted in his Executive Order. "But the experience of this and all other nations has demonstrated that some degree of censorship is essential in wartime, and we are at war."[9] Seven months later, the American Civil Liberties had signed on to the program, noting "Censorship arising out of war has raised almost no issues in the United States."

Price held censorship to the highest standard. The walls of his office had only one decoration—a handwritten paper quoting Owen Tweedy, British journalist and civil servant: "A censor needs the eye of a hawk, the memory of an elephant, the nose of a bloodhound, the heart of a lion, the vigilance of an owl, the voice of a dove, the sagacity of Solomon, the patience of Job, and the imperturbability of the Sphinx."[10]

Price was a perfect choice for the censorship job. He knew Americans disliked censorship, a concept that was fundamentally abhorrent to him as well. He was known for telling his censorship staff that in a popularity contest, censors would be at the bottom of the scale. While Price attempted to maintain a low profile for the Office of Censorship, eschewing public debate, popular discussions, an alphabet name, and a motto for the office, Washington pundits quickly circulated their own motto—"O.C. can you say?"

Before taking the censorship job, Price was a newspaper man, committed to informing the public, not withholding information. His journalism experience was invaluable to the office in monitoring without censoring newspaper publications and radio broadcasts, an arrangement that had not been practiced in other times of conflict. He knew key figures in the news business. As a former Associated Press correspondent, he traveled the country covering presidential candidates and political conventions. He moved up to the Associated Press bureau in Washington, D.C., and then became executive news editor in New York, a top position overseeing a massive, international news organization managing 1,400 news outlets throughout the country.

Price understood the federal government, key players, and idiosyncrasies of Washington politics, having lived in D.C. for twenty-two years covering federal affairs for AP. As a newsman, Price adhered to the mantra of political neutrality, a valuable characteristic in a censorship operation where political favoritism would be toxic.

Price had the power of a presidential mandate without the army of personnel to carry it out. He turned to the military, which already had trained censors, and absorbed them into his civilian operation. As Price began assembling a team of leaders and on-the-ground censors, he hired as his immediate assistant Theodore Koop, a former news editor who had worked for Price at the Associated Press office in Washington, D.C. At peak employment, the Office of Censorship operated out of eighteen censorship locations throughout the United States and territories with more than 10,000 civil service employees opening and examining nearly a million pieces of mail a week.[11] The censorship operation spread halfway around the world, with postal and cable censors in every major city, U.S. territory, and friendly countries such as Iceland and Mexico with an army of censors that grew to 15,000. Koop describes them as a cross-section of America, many with ethnic or foreign backgrounds invaluable in scanning messages in foreign languages.

Early in the war and the censorship initiative, the Army and Navy took control of military censorship, leaving the Office of Censorship to oversee civilian censorship. The Navy left some personnel in the Office of Censorship to monitor cable communications related to shipping. Mail was censored on each ship.

2.

A. K. Schnell Mus. 3/c
U. S. C. G.
U. S. Mn. Greely A. P. 141 T. Div.
% F. P. O
San Francisco, Calif.

The quarters where we are bunked is
in the rear part of the ship

so you can see it is pretty tight fit,
but this is heaven compared to a D. E.
or so the fellows tell us.

If I didn't miss my darling so much
this really wouldn't be too bad but as it
is I can't feel happy when you are
not with me. I can hardly wait till you
can come here sweetheart. I meant to tell you
our chief had his wife come out and she
is staying on unless we get assigned
another home port.

We already have a name, who gave it
to us we don't know but it is the Greely
Grenadiers. The photographer aboard is going
to take our picture. We sure are getting
publicity being the first rated C. G. Band
aboard ship.

I am afraid I will have to get busy
now sweetheart so will close. Honey I
realize more every day how perfect you
are and how much I really love you
and miss you.

                    Completely yours always
    mousie, hugs & kisses,

                    Artie (Artie the sailor)

The portion of this letter that referred to the sleeping quarters was cut by Navy censors.

With a flurry of encoded letters and uncensored letters delivered to the local Post Office, *Greely* musicians prepared for departure. Art wrote his mother in the last of his uncensored letters, "I will have one of the boys mail this when he goes on liberty tonite. I hope to write one more censored letter to you before we shove off." Letter writers crafted their personal letters with the knowledge that someone other than the recipient would read them. Some teased civilian and military censors by addressing them in letters. One wrote, "Dear Mr. Censor: I'm in love with this Leatherneck and I intend to remind him as often and as mushily as I possibly can. You have just got to wade through it. If you black out one single 'sweetheart' or 'apple dumpling,' I'll make your life miserable."[12] The "sweetheart" and "apple dumpling" got through, but the XXXs at the bottom were removed from this, and every love letter that passed through military and civilian censors' hands. Art addressed the censors a few times. Flo had sent him a joke and he considered sharing one of his own. "I could tell you some that I heard around here, but I am afraid the censors would cut them out for immoral reasons—right Censor?"

Censors sometimes became more than impartial observers, inserting their words or in one case, the censor himself, into the correspondence. An Army captain reported missing on a flight over Java called his wife to let her know he was safe. The connection was poor and when she failed to understand where he was, the censor chimed in that it was Malang, Java. The woman was relieved. A censor in the Seattle Post Office reviewed a letter from an elderly woman in ill health who was traveling to Seattle by ship expecting to meet the relatives who were the intended recipients. A letter with any information about shipping routes, departures, and arrivals would not get past a censor, and besides, the ship was due that day. The censor met her at the dock.[13]

One of Art's letters to his mother-in-law inspired the censor to interject a comment. Art wrote, "It is impossible to tell you where I am going." The censor wrote on the letter, "Right. Censor." Flo reported the few times his letters were censored with scissors, leaving a gaping hole in the text. One letter mailed from the ship began with a description, "The quarters where we are bunked is in the rear part of the ship." The censor's scissors edited a one-inch gap in the letter, which picked up with "so you can see it is a pretty tight fit..." The original was returned to Art for revision, removing "58 fellows slept in a room about the size of our living room hall and beds, some in bunks four high, with him in the bottom." He mailed the original letter through the Post Office.

Before the *Greely* left port, Art wrote that Navy censors had become zealots in their scrutiny. "It is hard to know what to write and what not to write. Hal Brody wrote a letter home and said, 'We are all getting orientated to this life.' The censors sent the letter back to him with a warning not to do it again or he would be restricted from writing for a month. He didn't even know what was wrong until they told him about the word 'orientated,' so you can see how fussy they are on this ship." While Navy censors scrutinized letters for compromising information, they also waded through the personal lives of sailors and spouses. Flo, for one was not happy about it. Art had written that his back was sore from sleeping on deck. "What I wouldn't give to be back in our bed, Honey—I would enjoy sleeping in it too—oh me. Chalk me up for 10 cartwheels." Flo replied, "I hope I never have to look that censor in the eye. Your letter must shock him, especially the part about the bed etc. Getting pretty brave, aren't you?"

The first voyage through the Pacific and Indian oceans to Asia would have been the trip of a lifetime for the musicians had it not been for the war. They observed places and people that most Americans had never experienced. Wishing to share these experiences by letter with their wives and loved ones was only natural. Descriptions of an ocean voyage, exotic places, different cultures, and people would be an enduring travelogue in peacetime. But the censors were as keen on eliminating stories of past experiences as they were about present and future destination.

While the Office of Censorship and Navy censors scrutinized letters, cables, and phone calls to restrict the release of information, the Office of War Information oversaw the promotion of information to cast the war in a favorable light. The strength and character of the American people, the heroes that inspired soldiers to fight, and the patriotic workers who produced war machinery were propaganda themes, while stories and pictures of grim war conditions and casualties were filed in the recesses of the office. It was a balancing act. Consider the difficulty of publicizing the heroics of Signalman First Class Douglas Munro. The Coast Guardsman awarded a Medal of Honor during the war was credited with saving Marines in the battle of Guadalcanal but died doing it. While stories from the battlefield were sanitized, stories of war, news accounts, and photographs pierced the Office of War Information veil of the "good war." Ernie Pyle reported on the war from the perspective of troops fighting on the front lines and brought their stories of war to the homeland. His columns described the lives of ordinary soldiers and their struggles. Pyle traveled with American troops to the battlefields of North Africa, Italy, Sicily, and the Pacific, covering the war in realistic terms while maintaining standards of self-censorship required by the Office of Censorship. His columns, carried by more than 400 daily newspapers and 300 weekly newspapers, told stories of determination and patriotism peppered with grim stories of brutality and death. "While military services censored the images from the battlefront and so provided a sanitized picture of the war, Pyle gave his readers a real sense of the brutality of the struggle. Simple sentences—such as 'The frontline soldier has lived like an animal for months and is a veteran of the cruel, fierce world of death'—summed up the relentless difficulties ordinary soldiers faced and brought them closer to loved ones at home."[14]

Publishing photographs of dead American soldiers was prohibited during World War I. News media accepted a voluntary prohibition of such photographs in World War II, although photographs of dead enemy troops abounded. Pictures of World War II POWs were rarely, if ever, shown. Twenty-one months into the war, photographer George Strock photographed American corpses on Buna Beach in New Guinea. The battle over publication of that photograph began with a captain and escalated to the office of the president, with Roosevelt giving approval of the publication in *Life Magazine*. The picture appeared in the September 1943 edition. Three corpses of U.S. soldiers lie on the beach as a small wave curls, poised to wet the sands that embraces them. The editorial accompanying the photograph concludes, "And so here it is. This is the reality that lies behind the names that come to rest at last on monuments in the leafy squares of busy American towns."[15]

While casting the war in a favorable light, the Office of War

Information was intent on instilling in all Americans the importance of self-censorship. The iconic statement "Loose Lips Might Sink Ships" was pasted on posters displayed on bases, ships, factories, and public buildings to remind everyone that the release of sensitive information could be dangerous in the wrong hands. Other posters showing a sinking ship bore the caption, "A careless word, a needless sinking." One showing a sailor and a woman embracing read, "Sailor beware! Loose talk can cost lives." Another portrayed a sailor carrying a sea bag over his shoulder. "If you tell where he's going... He may never get there." Dozens of others were commissioned by renowned artists and were carefully crafted to create fear, emotion, compliance.

The reckless spill of sensitive information was as much a risk with high-ranking officials as it was with enlisted soldiers and sailors. No one exemplified the dangers of loose lips more than Congressman Andrew May. The chairman of the House Military Affairs Committee visited Pearl Harbor after the attack and learned that the Japanese had no idea how deep U.S. submarines could dive and set the detonation of depth charges too shallow. May attended a news conference where he made public the flawed Japanese strategy, detonating an explosive reaction from Vice Admiral Charles A. Lockwood, commander of the U.S. submarine fleet in the Pacific. Lockwood later accused May and his loose lips of costing the Navy as many as ten submarines and 800 crewmen as the Japanese reset their depth charges to deeper water.[16] No direct link was found to connect May's statement to the loss of the submarines or that the Japanese modified their settings as a result of the statement. But May should have known better.

In late May 1944, Major General Henry Miller, commander of the Ninth Air Force Service Command in England, was shooting off his mouth at a London cocktail party. The Allies were preparing for the invasion of France at an undisclosed location on an undisclosed date. A cocktail party snitch—a woman among several who listened to the general—reported to security forces that he had said, "On my honor the invasion will take place before June 13." The information was relayed to General Dwight D. Eisenhower, who immediately slashed Miller's rank and sent him home. Censors at the highest level, fearful that if the story of the demotion got out, so too would compromising information, kept a lid on it until the invasion was well underway on June 6. Two days later, newspapers reported that Miller, a West Point graduate, classmate of Eisenhower and Lt. General Omar Bradley, and a long-time military officer with experience dating back to World War I, was relieved of his position for blaring top secret information at a cocktail party.[17]

CHAPTER SIXTEEN

War Letters

Glenn Miller, famed big band leader and director of the Army Air Force Orchestra, ended Saturday broadcasts with a request to "Keep those V-Mail letters flying to the boys overseas. Mail from home is number one on their hit parade. They're doing the fighting. You do the writing."[1] Lt. General James H. "Jimmy" Doolittle shared those priorities at a Glenn Miller concert in England in July 1944. He grabbed the microphone and addressed the crowd and Miller with the statement, "Next to a letter from home, Captain Miller, your organization is the greatest morale builder in the E.T.O. [European Theater of Operation]."[2]

Popular magazines such as the *Saturday Evening Post*, the *Woman's Home Companion*, and *House Beautiful* splashed front-page images of troops in various poses and places reading letters, the images accompanied by articles about the importance of war mail. Some magazines offered tips on letter writing while churches, schools, and community groups sponsored letter-writing campaigns. Dozens of popular songs during the war voiced the theme of war letters. Mail was considered the cornerstone of morale building, assuaging the physical and psychological wounds of those on the battlefield, and maintaining faith and hope for those at home.

The U.S. Post Office was a major component of the war effort, sorting and managing billions of correspondences between troops abroad and families at home. In its 1942 Annual Report, the Post Office noted that "frequent and rapid communication with parents, associates, and loved ones strengthens fortitude, enlivens patriotism, makes loneliness endurable, and inspires to even greater devotion the men and women who are carrying our fight far from home and friends. We know that the good effect of expeditious mail service on those of us at home is immeasurable."[3] In 1943, when Art began writing letters from the Manhattan Beach base during the early days of the war, the Post Office had delivered more than half a billion pieces of mail overseas. During the last year of the war, 3.5 billion pieces of overseas mail went to troops. In the same year, the Navy Postal Service processed 8 billion pieces of mail to and from sailors abroad, while the Army Postal Service processed 2.5 billion. The V-mail system used between 1942 and 1945 counted more than a billion photographic letters.[4]

Speed of delivery was an important factor in mail communication, but the initial system was not fast. When the United States entered the war, a letter from home took at least ten days and, in most cases, much longer to reach the troops in combat zones. By the end of the war, letters were arriving in seven or eight days and, in some cases, five days.[5]

The Navy recognized the value of letters and took steps to improve slow delivery. The *Sparkles* newsletter describes the revamped system: "When a man first arrives overseas on a ship his morale is at its lowest ebb. Mail from Mom, Dad, and Mary in the past didn't catch up with him for three weeks to three months after he left the States. Under the new system, mail will greet him on his arrival overseas, and the letters that just missed him as he left the States will be ready to be delivered at his ship's side when he first puts into port."[6]

Every sailor leaving the States was required to write his new address on a correspondence card for his commanding officer and his anticipated commanding officer on a ship or station. The new commanding officer knew in advance the sailor would be arriving and held his mail.[7] Sailors learned what to expect from the delivery system. They wrote their letters in anticipation of mailing them at ports, where they hoped to receive some. They learned to read with joy and enthusiasm about events that had happened days or weeks ago. They wrote even when letters would arrive after a reunion, as if the substance of words on a page made permanent the affection and emotion of ethereal conversation. Art wrote such a letter to Flo from the *Greely* after it docked in New York, expecting to see her before the letter arrived. When the *Greely* docked in Calcutta, the lack of letters left the musicians in a state of malaise. But when the *Greely* arrived in Suez, Egypt, June 8, the mail had caught up with them—six bags of mail for the crew, including five letters for Art.

Through letters, Art and Flo, as well as many other war couples devised plans for a future without war. From early recollections, I understood my mother was reluctant to have children until the war was over and Art had survived. She had reasons. Flo's elementary music pupils were as wild as jungle kids, climbing walls and making animal noises. A musical duet in a school assembly bombed when one young musician broke his clarinet, and Flo, red-faced, tried to fix it while the audience fidgeted. Pupils she supervised in the summer playground job were sweaty, smelly, and noisy. Francine's baby screamed and wet the bed. Others her age were cooped up in a small, noisy room with a babysitter. "It's one argument against taking fathers into the service," she wrote. "And on the other are the mothers who work just because they want more money and can't be bothered having their children around." In prewar letters, when the couple taught in separate schools, there was no mention of having a family. Art may have initiated the discussion from Manhattan Beach base when he floated a trial balloon during basic training, noting that the military regimen prepared him to be a father. Plans became more detailed when he was transferred to California and the *Greely* when letters traversed half a globe, and days of waiting for them seemed like years. Flo began shedding her aversion to children as she envisioned a life in the country after the war. "If I could only get away from this place—New York, I mean. Honey, I don't think you'll have a bit of trouble with me after the war. I'll be docile and content to live in the country. Besides, I saw the cutest baby today. Very young too. Maybe they're not so bad at that age after all." Then Francine's baby outgrew bedwetting and began making sense. "She speaks so plainly but still baby like, and in her prayers she said God bless Florence and Artie and Minnie and Florence's Daddy and big Fred boy. It's marvelous how she keeps talking about them—remembers everything. Honey, hurry home so we can have one. I'm not kidding this time." With the end of the war in sight, the plans were formalized in letters. Art wrote, "Remember last year when we thought the war would be over by now? Wish I could see the end of it in sight, but it can't last forever. We have that family to start though, dearest and that can't wait forever." Flo put the finishing touches on future plans when she suggested they get a sailboat. I imagine Art's joy.

Self-censorship

While soldiers and sailors groused about the restrictions over their letter writing, many adopted their own standards of self-censorship to put a good face on their tiny sphere of war. They did not want their letters to cause stress or anxiety at home. Kimberly Guise, with The National WWII Museum, reviewed thousands of letters troops sent to loved ones. She found many devoid of words of war, reflecting instead routine or upbeat messages with telltale omissions.

Sometimes, it is not only what is on the page that is important, it is what the writer could not or did not want to convey that is truly meaningful. Many wartime correspondents did not want to create anxiety for those back home or for those serving. They tell of the everyday, they tell of ordinary meals eaten, letters received, and inquire about the health and safety of faraway loved ones. Rarely will they expose sentiments of fear, despair, or elements of danger. Only by reading between the lines can readers judge the true meaning or historical significance.[8]

Art downplayed even the most insignificant news about adverse conditions. After he wrote from the Manhattan Beach base that he had a cold, he attempted to mitigate the negative news. When the platoon was required to hang their wet laundry around their bed, Flo became concerned that the practice aggravated his cough. He wrote, "Sweetheart, don't worry about my cough as it is entirely gone, honestly. I don't hang my clothes around the bed only when it rains." From the ship, he wrote, "I'm feeling fine, honey. Maybe this ship life agrees with me." After arriving at Melbourne, the *Greely*'s first foreign port, he wrote, "I am feeling fine, sweetheart and everything is going along nicely." He wrote his mother, "Don't worry about me, Mom. I never felt better in my life. By 10 o'clock I am ready to sleep, this sea air makes me awfully sleepy."

His letters from San Francisco to his mother were so optimistic and happy that Flo demanded to know why. He explained, "Darling, the reason that my mother mentioned that I felt happy is because I have written her really cheerful letters so she won't be brought down." The same strategy may explain his promoting Hawaii as a credible destination after he was assigned to the *Greely*. Flo reacted favorably to destination Hawaii, relating that Fran, her best friend's mother, would like Art to bring back a grass skirt for daughter Francine.

It was acceptable to share minor injuries and setbacks to add humor and to create the impression that letters were constructed with candor. Art mentioned a minor incident on the Manhattan Beach base when he burned his hand trying to light a cigarette. The entire book of matches blew up in his hand, burning his thumb. "It's going to hurt to hold that T-bone for colors tonight."

The writing paper and handwriting in the exchange of letters tell a story of unique situations war couples faced. Most of Art's letters were written on fine, fairly heavy stationary at times with letterheads denoting the U.S. Navy, the USO, Red Cross, or other organizations that recognized the important letter links between those away in service and those at home. Most of his handwriting was meticulous, each word carefully penned, most sentences in straight lines. At Manhattan Beach, he wrote between events during a busy schedule, sometimes closing with a "lights out" notification or the need to leave for rehearsal. On the ship with free time and access to a music room with a desk, he developed a letter management plan that involved a filing system and a rereading of his wife's letters. "You know, honey, I have quite a complicated filing system for your letters. After I receive your letters and read them over a few times, I put them in a large manila envelope, then I get them out again and read them over before answering; after that, I file them away according to their postmarks."

In the turmoil and uncertainty of war, stories of the past reaffirmed and strengthened the relationship in separation. Art reminded Flo of the time he skipped band rehearsal in college to be with her. "You know honey I was lying out on the deck tonight watching the stars, I couldn't help but remember the time I skipped rehearsal at college and we sat in the back yard... and talked about everything under the sun while watching the stars." She wrote him about how sitting next to man with a box of baby chicks swept her back to a scene in the Catskills. "Remember the cute chickens we hatched in Youngsville with our names on the eggs? I still think you bribed the hen to sit on yours more than mine." She recalled that incident in college when

Art "sneaked out for a beer when my chair was singing (worm!)." Amid the euphoria of receiving a letter, Flo mused about the downside of war separation, regardless of the Navy's efficiency at managing mail. "Your letters help a little bit, each page is full of charms. But Darling they aren't quite enough. For letters don't have arms."

Flo carved out time for letters in her busy schedule, often writing on the train traveling between her apartment and work or at the station waiting for a train. Writing letters was a conditioned reflex triggered by boarding a train. "On the train again," she wrote. "I spent half my life on these blamed trains—honest." The swaying, clicking and clacking of the bumpy tracks were not conducive to penmanship. "I don't know how the poor thing runs on them, do you?" she asked. "I hope you can read this—all I do is hold the pen still, wait for a jolt and that makes another word." Many of her letters were written on thin paper—cheap to buy and cheap to air mail—while others were torn from a small, three-hold folder, the edges gouged. She preferred to write in ink but at times found herself on the train without a pen. "I'm a tramp to keep writing you in pencil, but it's better than that child not getting a letter." Her handwriting seemed ragged, her lines, slanted, her letters not always clear. She mistakenly opened one of her own letters and was surprised at the penmanship. "Before I realized what I was doing, I thought your handwriting looked a little sloppy."

The differences in their letters reflected their disparate lives determined and defined by war. Flo assumed the responsibility of managing the household, fixing the dresser, fixing the car, maybe selling the car, teaching full time, giving private lessons, rehearsing the glee club, fighting the taxi driver, and devising a plan to get to San Francisco. Those activities and full-time teaching took up most of her time. After a stressful day, she would come home to the apartment she had shared with Art, who had been replaced by Genny, who had become increasingly difficult to live with. Some nights, she'd crawl into bed exhausted, begin a letter, and finish it the next morning.

Before shipping out, Art was enjoying San Francisco, with its ethnic restaurants, nightclubs, and musical performances almost daily. Musicians were spared some of the loading duty to play for the work crew. Once aboard, they enjoyed three meals a day, no cooking, and no dishes to wash.

They picked up laundry on Fridays, cleaned and folded, and slept in special quarters. Art was among his Manhattan Beach band buddies and bunked next to the much-admired trombonist Eli Bublick, who entertained him with stories. He wrote letters during free time at a desk in the band room, rereading them before sending, sometimes inserting a missing word with a caret. In the absence of a torpedo attack, it was a pretty good life, as portrayed in letters.

Sailing through the beautiful Pacific and Indian oceans, mostly in fair weather, on a new ship, to the music of the *Greely Grenadiers*, would be a dream cruise. The bands played on deck, the military band blaring patriotic marches, the dance band flooding the deck with hits of 1944 and 1945, romantic dance tunes sure to pique the imaginations of crew and troops. While the bands performed, the crew shot craps and played cards, watched flying fish skim the ocean, and attended movies nightly. The chow lines were long, but the food was acceptable. The night sky was as vast as it gets with water horizons on all sides and a million points of light strewn over the black backdrop of night. Men slept under those stars on deck where the air was fresh and cool. They watched a whale breach and spout, and were dazzled by multiple rainbows in a clear sky. Spectacular sunsets were an evening attraction. Those idyllic scenes, described in dozens of letters from the ship during its first voyage, surprised me. As a kid, I heard harrowing stories of torpedoes crossing the bow, a typhoon, strategies for abandoning a flaming ship, and strategies for survival in a lifeboat. Most stories would not have been targets of censors' scissors. They were omitted through a "don't worry about me" self-censorship mentality practiced by sailors and soldiers in some of the most hostile and dangerous conflicts of the war.

Art's letters did not describe strategies for survival that sailors packed in their mental duffel bags to mitigate the fear that, at any second, their ocean cruise could blow into an inferno, sending survivors scrambling for the safety of the water while others trapped below deck would be lost. Surely, most had some plan for survival, however fantastic in the event of a torpedo attack. My father would be seated at his station in front of dozens of valves waiting for the phone call to indicate which section of the ship to protect with carbon dioxide, assuming an officer near a phone survived to give the order, the communication system was intact, the whole ship was not blown apart, and my father was alive. Each person on the ship received a copy of the booklet *How to Abandon Ship*. The booklet began with a section on preparation and training before embarking, and in the absence of that, reading the booklet en route. Abandon ship drills, lifeboat drills, what to take, what not to take, how to rig a sail on a lifeboat, and other practices were detailed in the booklet, which was peppered with stories of sailors who did it right and survived and sailors who didn't and died.

My father's war stories reflected the survival strategy portrayed in *How to Abandon Ship*. The *Greely* crew trained in San Francisco to prepare for a torpedo attack that created a sea of burning fuel. In a swimming pool, Navy trainers simulated the experience. Trainees dived into the fuel, swam under the flames as far as possible, emerged with hands extended together above their heads, and parted the flames. After taking a lungful of air, they submerged again, swam as far as possible, surfaced, parted the flames, gasped air, and dived, repeating the maneuver until reaching the boundaries of the inferno. As a young teenager, I swam underwater at the Van Hornesville Central School swimming pool, looking up at the undulating blue sky and seeing, instead, dancing orange flames. I would swim up, part the flames, take a deep breath of clean, Upstate New York air, and go back under to the safety of the water. Sure, I could do that, I was a good swimmer. I didn't know about toxic petroleum fumes and what they would do to a swimmer's lungs. I didn't know that the *Greely* carried up to 558,600 gallons of fuel, enough to set much of San Francisco Bay ablaze. Without that knowledge, Naval warfare seemed exciting. As an adult, I realize that such training was intended to offer sailors a nugget of optimism as they sailed hostile oceans with a nagging feeling of utter helplessness in the event of an attack. Such stories were not suitable for sharing with loved ones at home while the war was on.

This booklet was distributed to *Greely* passengers before departing for hostile waters.

The Navy's *How to Abandon Ship* booklet took a dim view of surviving a burning ship by swimming under the flames, although the booklet cited a few success stories and offered some recommendations. Fumes from exploded torpedoes were toxic but with a wet handkerchief treated with a few drops of aromatic spirits of ammonia tied over the mouth and nose, the toxicity would be reduced. The booklet recounts the story of an oiler named Anderson on the tanker Harry F. Sinclair, torpedoed April 11, 1942, off Cape Lookout, North Carolina. He swam under the flames until he reached clear water. "Few men could have accomplished this," the authors noted.[9] A messboy in the same tanker surfaced, fought back the flames, took a breath, and swam underwater again until clear of flames, as Art was instructed to do. He survived, severely burned, but if he had a woman's bathing cap over his head and ears, fastened under his chin, his burns would have been less severe, the booklet noted. I did not find any reference in the letters or ship's documents regarding provisioning with bathing caps or ammonia, although sailors could have purchased their own had they received the booklet before boarding. Surely with fifty women on board, bathing caps were available but in short supply. Finding them amid the chaos of an attack would be tricky, and fighting the women for them would not be chivalrous. The cork life preservers sailors and crew were assigned would make survival by diving below the flames futile, the booklet warned, and neither of the survivors of flames would have made it had they been wearing them. No flaming ships or flaming seas were mentioned in Art's letters.

Musician Roger Hartman, who complained of chronic kidney problems in Manhattan Beach and San Francisco, kept Hostess Twinkies in a safe place in case the *Greely* was hit, and he managed to get into a lifeboat. Hartman would sit with my father on deck with a map and estimate the ship's position. Then he would find what he believed to be the nearest friendly island. As they cruised through open water, he often related his survival strategy. "If we get hit, Art, we'll go for New Caledonia" or some other haven with his bag of goodies. Surviving a torpedo attack, the fire, the explosions, and making it to a lifeboat was no guarantee of survival. Hoarding Hostess Twinkies amid a crew of hungry sailors on a lifeboat could get contentious. Further, the sea is as cruel an adversary as an enemy submarine. Surely every sailor had heard of the disaster two years earlier, when the five Sullivan brothers from Waterloo, Iowa, were serving on the U.S.S. *Juneau*, a light cruiser off Guadalcanal. She retired from battle after a torpedo caused severe damage. Barely able to make headway, she was an easy target. The next day, she was hit again with a torpedo that detonated the explosives on board. The ship was blown apart and sank in less than a minute. The explosion was witnessed by the crew of the U.S.S. *Helena*, which reported to a B-17 that the ship was lost and survivors were in the water. The message, however, never reached the command center. One military account[10] contends four of the Sullivan brothers failed to make it topside to abandon ship. George Sullivan, wounded during the previous attack, abandoned ship and pulled himself onto a boat. Another account notes that Tom Sullivan also survived the explosion but was seriously wounded and died shortly after the ship was lost. Five days after the Juneau was destroyed, George Sullivan, mad with grief, took off his shirt, went into the water and was not seen again, possibly a victim of sharks that had killed many of the sailors who made it off the ship. Nine days after the *Juneau* sank, a rescue team located ten survivors. An estimated 140 sailors had abandoned the *Juneau*.[11]

An ocean storm could be fearsome and dangerous, even for those on a massive troop transport ship. My father told me he endured a typhoon sitting in the stern of the ship, ducking sheets of water flying from the bow over the length of the ship. The familiar throb of the engines would rev up to a scream as the props left the water while the ship seesawed on the crest of a wave, then dropped to a drone as the bow plunged into the trough and rose to meet the next wave. Because an attack during severe weather was unlikely, sailors found relief from anxiety amid the storm even if it meant crashing in the music room to endure sea sickness. A lead ship in the convoy was so badly damaged that it put into Australia for repairs. The ship that took its place was destroyed by a torpedo. None of these stories was in his letters. It sure looked like smooth sailing for the *Greely Grenadiers*.

As the musicians wrote letters almost daily to loved ones, they began running out of words and phrases. How many ways can you say, "I love you," "I miss you," and "I am yours alone" in letters the writer knew would sit in the hold of the ship? "Until we get mail, hon, I will have to cut down to every other day. There just isn't enough to write. I read the first page of this letter to Joe Perna, and he said it was practically the same thing that he said to May."

After departing Ceylon in May, Art and Joe Perna discussed the downside of letter writing from the ship, where boredom had set in, news of exotic ports was off limits, and words of love had become cliches. For weeks, they heard nothing from home and wrote in the emotional void, hoping that at the next port, a stack of letters awaited them.

The *Greely* wives also were dealing with a void in correspondence while volumes of letters awaited the *Greely Grenadiers* at the next port. "I just heard Flip [Manne] got a cablegram from Shelly and that he's receiving mail," Flo wrote. "We were all pleased to hear that.... It's good to know that somebody is because we certainly aren't. But we consoled each other, saying we would soon."

In May 1942, the escort cutter *Icarus* sank U-352 off the North Carolina coast, capturing America's first enemy combatants of World War II. Over the course of the war, Coast Guard-manned warships sank eleven U-boats.

CHAPTER SEVENTEEN

Coming Home

May 27, 1945. Dear Darling

How's everything in my life today, in other words, how are you sweetheart? This afternoon I fell asleep on deck and I had my wallet opened to our picture. I'm still wondering how many people passed by and looked at the picture, because finally Eli Bublick woke me up and asked me what I was doing sleeping with a picture in my hands. I tried writing mom a letter tonite, but there is so little I could tell her it was just an irritation to write. I can't tell you how good it seems to be on our way again, at least we have a cool breeze now. It's only a matter of a few days till my birthday, honey, 27 years old—that seems kind of old to me. I'm awfully happy that we reached our destination and have started back. You sure will be hearing a lot of talking from me when I get back. All the fellows are asking one another what they can write home about and it isn't funny because there is so little we can say.

The *Greely* sailed for the East Coast with a load of war-weary Army soldiers, Army nurses, Red Cross women, and the ship's crew. "We have certainly covered an awful lot of ocean in the past month and a half. It's nice but I'll still take America with open arms," he wrote, hoping he'd be allowed off the ship with time to spend with his wife and a few days retreat to the Adirondacks. Until then, he had to contend with personality issues on the ship. Conflict with trombonist Jesse Ralph had been building during the voyage. A month earlier Jesse rehearsed with a brass quartet for a church service and didn't show up for the service. Art was a last-minute stand-in, and had not seen the music. He and Jesse had not been on speaking terms since even though they played together in the military band, that is for the sessions Jesse Ralph attended. He missed half of them. Art replaced Jesse in the brass quartet for church services, sick bay, and special occasions. He was not happy about playing in the stifling heat below deck, soaking his shirt with sweat. Finally, Jesse Ralph was kicked out of the entertainment department and assigned to another department on the ship. At the same time, Art's relations with Frank Boeberitz warmed when Frank gave Art his ration of two beers and fixed up his trombone case, which was falling apart. "I guess he is a pretty good guy after all. I just don't understand him too well," Art conceded.

No one expected more letters from home until the *Greely* reached the States. Musicians were feeling lost and lonely in the absence of communication from wives after only a few months' absence. The troops the *Greely* was bringing home had a much greater void to deal with. The void and the boredom were fertile grounds for marital doubts. What or who would they find when they got home? But Art maintained his confidence. "Did I ever tell you I never got jealous of you, honey? You know I could very easily be jealous, but I know my sweetheart would never give me any reason to be but still I am a little jealous just because I can't be with you."

The trauma of war was evident in some passengers on the *Greely*. Just before 1 p.m. June 5, a lookout on the stern reported a man overboard, prompting a general quarters order. Crew tossed a life ring and readied a boat for rescue while the *Greely* circled back. As the ship maneuvered near the man, crew lowered the boat, and in just over fifteen minutes rescued Sidney Dillman, identified as a mental patient. The ship's senior medical officer examined him and found no lasting effects from the submersion but prescribed no treatment for post-traumatic stress. The next day, passenger Chih Tsi Hsien died suddenly from endocarditis, unrelated to war. He was given a Navy ceremony and buried in the Red Sea.[1] In an effort to maintain morale, the *Greely* held a birthday party for everyone with a birthday in the first few days of June. Art's was June 2, but he slept through the celebration and missed out on cake. But the *Greely* was making progress on the voyage home, location encrypted.

June 8, 1945. Dear Darling
I sure had a very pleasant surprise today. They unexpectedly picked up about six bags of mail so we got those long-awaited letters. Please

excuse the writing, darling, but you can't imagine my excitement at getting those letters. I'm zowied if you know what I mean and just completely knocked out, it was so wonderful hearing from you. I can honestly say that I had given up hope of receiving any mail till we get back to the States. In all I received five letters from you postmarked between May 17 and May 23, one from Mom and [brother] Bill, and one from good old Jarlath Perkins. It never dawned on me that the mail would also leave the ship from here or I would have gotten the first part of this letter June 6 finished. [The code ends without completing spelling of "canal."]

Just when some musicians had given up hope of getting letters from families, the mail caught up with the *Greely* as it transited the Suez Canal and docked at Port Said in Egypt—nine sacks of precious, long-awaited letters for crew and passengers.[2] Unprepared for access to postal service, musicians dashed out letters for return mail with no indication they would be delivered before the *Greely* reached port. Flo deciphered the code of the earlier East Coast Dear Darling letter and wrote that the *Greely* wives had abandoned plans to go to the West Coast. Flo used the code to request an East Coast arrival date. "Please try to remember when aunt Matilda's birthday is. I'm sure it must be in early July."

During boring days at sea, bridge was the card game of choice for many musicians and crew. Art and others had been playing bridge almost daily since the *Greely* left California, and their enthusiasm was infectious. As the *Greely* sailed through the Arabian Sea, the game had become so popular that many with and without experience wanted to join. Officers and passengers hung around the bridge tables hoping to learn the game. In one round, Art and his partner played separately with a major, a captain, and a lieutenant. His team won three out of four rubbers and missed a grand slam by one trick. On another occasion, he and an experienced bridge partner invited Chaplain Hugh Miller and a partner to play bridge. Miller was not experienced and considered a few cents a point a harmless wager. But when the game was over Miller and his partner had lost far more than they had expected, and Miller felt he'd been taken for a ride. A man committed to the morale of his troops, he did not hold a grudge. After honing their skills

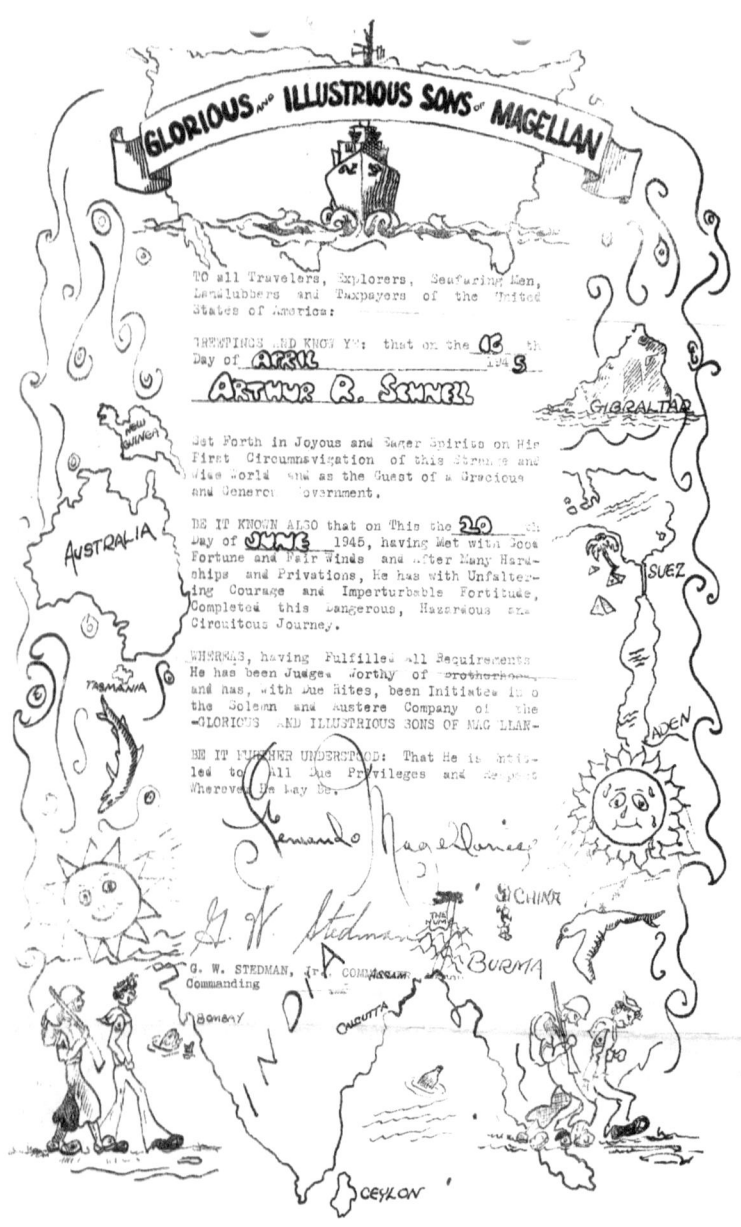

Arthur Schnell and other *Greely* musicians were honored as Glorious and Illustrious Sons of Magellan after circumnavigating the Earth on the first voyage.

during many hours and days of boredom, the officers arranged a bridge tournament for June 13. Art and Eli Bublick were partners. In the first table, they took on Belmont Ketchel and music Chief William Schallen, trouncing them. They did not make it to the finals. A Chinese team took first place.[3]

The *Greely*'s arrival June 22 in Newport News, Virginia, was celebrated on deck with an elaborate circumnavigation ceremony and a performance of marches and popular romantic songs by the *Greely Grenadiers*. It was time to honor the crew members who had completed a circumnavigation and were entitled to join the Glorious and Illustrious Sons of Magellan— if they could prove it. In step with marching tunes by the *Greely Grenadiers*, hordes of crew and passengers paraded around the deck, arriving at the Captain's bridge, where they crowded the area between Hatches 4 and 5. Commander Stedman began the rites by welcoming Magellan, dressed in the garb of old-time mariners. Magellan warned that membership belonged only to those who proved they sailed around the world, "facing many hardships with unaltering courage and imperturbable fortitude."[4] Magellan required that a representative of each of the armed services aboard relate a story that proved they had made the voyage. After each story, he deferred to Commander Stedman, who like a Roman Caesar in the Colosseum, faced the crowd of sailors perched on anything horizontal, soliciting a thumbs up or thumbs down for each candidate. With a rough count of the thumbs up and an assessment of the volume of cheers and hoots, Magellan announced the winners.

Bob Clements, water tender first class of the *Greely* crew, took top honors with his Baron Munchausen-style stories of impossible exploits and exaggerated heroics. Munchausen, a fictional character in 18th Century German literature, rode a cannonball, fought a forty-foot crocodile, and traveled to the moon. Clements told Stedman he sailed on a ship so large that the crew in the crow's nest required oxygen masks and descended with a parachute while crew in the engine room used an elevator to arrive on deck. Lt. Margaret Horton, of the Army Nurse Corps, told of treating a member of Merrill's Marauders, unconscious yet fit enough to continue marauding while Horton practiced self-defense. Clements and Horton received the most enthusiastic audience reaction. Magellan—later revealed as Rev. Gustaaf Swerd of the Office of War Information—presented Commander

Stedman with an honorary certificate of membership in the Glorious and Illustrious Sons of Magellan, having completed his tenth circumnavigation. Stedman also received a hand-painted plaque from the crew and passengers followed by a chorus of thanks for a pleasant voyage. Red Cross women passed out Sons of Magellan certificates to the *Greely* crew. The ceremony was repeated in sick bay. Art's certificate, intricately decorated with scenes of the *Greely's* travel, is preserved in an album stored in an attic in New York.[5] His circumnavigation story, if he had one, was lost as are details of the next six days he spent in Brooklyn with his wife. The band members and crew of the *Greely* received American and Asiatic-Pacific campaign ribbons. Many in the band believed these ribbons were tickets to shore duty for the remainder of the war.

Without enough time to get to Rochester to see his mother, Art summarized the last phase of his circumnavigation in a letter. He would do a better job on the travelogue another time. Back in the States, he caught a train to New York, met his wife, and crammed talk of travel between tennis games, swimming, and dining out—three days that flew by like minutes. Then he headed back to Virginia in the "most miserable train ride I ever had." The train was full and cramped, with Frank Boeberitz "heaving his cookies." After arriving at Cape Charles, Virginia, at 6 a.m. they took a ferry and a taxi to the ship, where twenty letters from Flo awaited him. He prepared for another sea voyage while reminiscing about the brief time with his wife.

"Words can't tell how much I enjoyed being home with you. In fact sweetheart I fell in love with you all over again, more so than ever before if that is possible."

CHAPTER EIGHTEEN

France and the Pacific

The *Greely* departed Virginia with musicians again scratching out letters. Art was on library watch June 29, his letter writing interrupted by people checking books in and out. He opened five of Flo's letters—sent before the reunion—and a bunch of letters from friends and relatives, wondering how he would have time to answer them all. Frank Boeberitz stopped by and asked if Art was trying to outdo him in letter writing—always a concern as their spouses knew which one got letters, when, and how many. Art distributed notes that Flo had enclosed for Tom Stokes and George Bland. After Danny Cowan's New York reunion with his wife, he related to Art how much Natalie enjoyed Flo's company. Art was not surprised. "All I can say is 'naturally,'" he wrote from the ship. "Still can't stop raving about how wonderful you look and what a good time I had at home. Pardon me but I guess I am awfully much in love with a certain sweet girl."

There was little time for love. The *Greely* was steaming eastward without troops. The crew, including musicians were assigned to painting the decks. The bands rehearsed for a shipload of troops expected on the return from Europe. After their brief liberty and reunion with their wives, some, including George Bland, looked pale and had lost their fervor for music. Perhaps they were not inspired playing to a nearly empty ship. The first dance band rehearsal was "nothing short of horrible. [Harold] Sachs is playing tenor sax and frankly it stinks." The show went on for a small audience despite the ragged performance. A few new musicians joined the bands. Only one, named Rocky, was good enough to play in the dance band, as far as Art was concerned. "The other fellows were dragged right out of boot camp by Chief Mulder, and they are not good enough to play in my high school band," he wrote. The dance band played in the mess hall and the musicians were rewarded with a ration of beer. Art managed to get a couple of extras from those who did not appreciate the brew. Later, some of the boys, including Chief William Schallen, got together in the music room for a "real, old fashioned bull session," with Eli Bublick and Harold Brody entertaining with "shady stories. We all had a million laughs." Art left the bull session early to write Flo a letter from his new quarters. The musicians had been moved from the stern of the ship to more spacious accommodations near the bow, where they had an elaborate music room for their instruments. As the *Greely* plowed eastward into building seas, the ship pitched as the bow rose and fell with the seas. About 1,300 miles out of New York, Art learned of the destination.

July 1, 1945. Dear Darling
I'm having plenty of competition trying to write this letter as Eli Bublick is practicing his horn in here. Yours always and Eli didn't make out so good in bridge tonight, honey, David Balogh and the editor of *The Salt* beat us by about 2,000 points. I'm very lonesome for you darling, I just finished opening your cute anniversary card with the letter inside. It's really lovely and especially the letter, words can't express how much you mean to me sweetheart and I will always be just the way you would want me to be. The eyes burn a little, I guess it's because you are so perfect in every way and so sweet in everything you say and do.

With the *Greely*'s arrival at the French port of Le Havre on July 7, 1945, the musicians saw the massive destruction of war. More than a year earlier, the Allies launched Operation Overlord, taking control of Normandy and moving inland to liberate Paris a few months later. Not far from Normandy, German troops in Le Havre, helplessly surrounded by British troops but heavily fortified, refused to surrender. On September 6, 1944, 348 British bombers unleashed a massive barrage of conventional bombs and firebombs over the southwest part of the city. The next day, six waves of Allied bombers targeted the eastern part of the city with a similar assault of conventional bombs and firebombs. Le Havre was blasted and burned to rubble. Docks and port facilities were damaged or destroyed. The destruction of German resistance took a heavy toll on civilians. More than 5,000 were killed and as many as 80,000 were homeless. By the middle of September 1944, the

United States Army began using Le Havre as a port to debark troops and supplies to support the Allies fighting their way toward Germany.[1]

By the time the *Greely* arrived eleven months after the bombing, U.S. troops swarmed the remains of the port city, having completed their mission to end the European war two months earlier. The *Greely* would bring some of them home. On July 9, the *Greely* embarked 835 troops and 43 officers, completing embarkation the next day with 2,631 enlisted men, 147 officers, two Red Cross women, and one civilian woman.[2]

Before the *Greely* arrived at Le Havre, Art had learned of the stateside destination and was eager to give Flo advanced notice in a coded letter mailed from the French port.

July 5, 1945. Dear Darling

It's now after taps, darling, and time all good men are in their sacks, but not me 'cause I want to get this letter off at our next port. Old Eli and I played two rubbers of bridge tonite for $.50 a rubber and we each won one rubber so financially it was a flop. You wouldn't believe how cool it is out here, it's hard to imagine that just a month ago, we were roasting. If you were out here you would know why the water is so cold at Riis Park [a seaside park on the Rockaway Peninsula in Queens]. Good old Angie [Bigatel] is playing third trombone in the dance band only he is using his euphonium [a valve instrument], it will work out OK too. We rehearsed the military band this morning on the "Malagueña" [originally a movement in Cuban composer Ernesto Lecuona's Suite Andalucía and later popularized with Spanish lyrics], William Tell Overture ("Memories"), I still cannot play that part. I know it sounds unbelievable, but I wrote your mom, [brother] Fred and my mom also tonight. Do you know we forgot to send Aunt Min a card and her birthday was June 18th...

The *Greely* wives' hopes the ship would dock in New York finally would come to fruition, and Flo received an arrival date. As the couple had no relative named "Aunt Min" and June 18 had passed, the coded arrival date would be July 18. Forgetting he was not to suggest a location and as yet unaware of the destruction of Le Havre, Art asked if she needed anything he might pick up when he arrived at the next port. He asked her to get some anchovies for him as an evening snack when he returned. Censors missed the indiscretion.

The *Greely* made her way into New York Harbor July 18. By noon she docked at Pier 88 on the Hudson River and began debarking troops into Manhattan. Pier 88 was a large port facility originally designed to handle cruise ships and well suited to large military ships. The largest, most elegant of cruise ships—the S.S. *Normandie*—docked there before the United States entered the war, her brief and tragic history a reflection of the intersection of leisure and war. Designed at the height of cruise-ship technology and culture, she reflected French pride in shipbuilding and elegance. First-class passengers enjoyed spacious and luxurious rooms decorated in the trendiest Art Deco style. She was the largest and fastest cruise ship ever built and was among the first to be equipped with radar for safety. When Germany occupied France in 1940, the Normandie was docked at Pier 88. The U.S. government considered the Vichy government of Germany-occupied France an enemy and seized the ship to convert it for use as a troop transport, naming it the U.S.S. *Lafayette*. In February 1942 while the *Normandie* was undergoing renovations for war, a welder ignited flammable lifesaving gear. With the ship's firefighting system disabled, fire broke out and spread. When the New York Fire Department arrived, firefighters got bad news—the French fittings for fire suppression on the *Normandie/Lafayette* were not compatible with their fire hoses. The ship quickly became an inferno, the spectacle attracting huge crowds. While firefighting boats attacked the flames from the port side, dockside hoses were deployed on the starboard side. Unfortunately, fireboats pumped far more water than dockside hoses. The *Normandie/Lafayette* began listing to port and taking on seawater through open portholes. She sank to the bottom of the Hudson River in her berth. By the time the *Greely* arrived, the *Normandie/Lafayette* had been raised, righted, and scrapped.[3] Other luxury liners were recruited for the war effort. The *Queen Elizabeth* and the *Queen Mary* both were converted to troop transports, prized for their high speed and enormous capacity. Like the *Greely*, they would bring home many thousands of troops.

Art expected an immediate reunion during the *Greely*'s ten-day repose at Pier 88. He wrote from the ship July 22 that he hoped to see Flo the next day, musing how strange it seemed to write a letter that would arrive after their anticipated reunion. He and the other musicians whose wives were in New York were disappointed to learn they would not get off the ship that day. A couple days later, Flo was experiencing car trouble as the engine overheated. Some men checked the car and found a ragged fan belt, which they replaced for $4, plus the cost of a couple beers—pretty expensive, she wrote, but she was able to drive it to lower Manhattan and cruise by the piers. With no word from Art, she tried to identify a large Navy ship docked in Pier 88. It was an unusually dark night and she could not see a name on the vessel. Based on the few sketches she had seen, it appeared to be the *Greely*. The site made her lonesome. It was frustrating to visit Pier 88 knowing Art was not far off in a cell of steel, as inaccessible as if he were in the Indian Ocean, their lives connected only by letters that took a circuitous route to bridge a gap of only a few hundred feet.

Resigned to remaining on the ship, Art looked forward to the next opportunity for a reunion. "I didn't want to get your hopes up about getting off before it leaves because you see we didn't, but wait till September, Bunny. September, Bunny? What have we planned for then? Oh yes, I know."

In a surprising reversal, the crew were given a few days' liberty, enough for cozy times with their wives but not enough for Art to visit his mother in Rochester. He later wrote to his mother, "Had I known we were going to have as much time off as we did, I would have insisted on your coming down. I couldn't have very well left Flo and come to Rochester alone and I know you wouldn't have expected me to do it. Frankly, I'd give a month's pay to be sitting on your front porch in that glider." Eager to enjoy the New York scene, some *Greely* crew flaunted military restrictions to have some land-based fun. Seven got permission to watch a show on the dock and ended up in Uptown; a dozen failed to return to the ship as required, some as late as two days. Others failed to get up or report to their duty station. One was assaulted in a barroom fight.[4] After the ship cleared New York Harbor, Commander Stedman convened a captain's mast[5] and imposed sentences involving extra work.

The *Greely* began embarking troops July 26 for the Pacific war. An advanced guard of forty-five officers, and 550 Army troops came aboard, followed the next day by 128 Army officers, twelve Navy officers, twenty-three Chinese officers, three British troops, one British sailor, fourteen civilians, 2,256 Army troops, and twenty-five sailors.[6]

Sailing down the Hudson River, the *Greely* crew got a look at the sights of Manhattan, including the Empire State Building. A few hours later a B-25 Mitchell bomber flying in thick fog was diverted from La Guardia to Newark, New Jersey, a route that took the plane over Manhattan. Pilots were flying slowly at low altitude, seeking visibility, when the Chrysler Building abruptly appeared out of the fog in front of the plane. The pilots swerved in time to avoid it, but the new course took the plane into the north side of the Empire State Building near the 79th floor. The fuel exploded, sending flames into the interior of the building. An engine crashed through the building and landed on a nearby penthouse apartment. Debris settled on top of adjacent buildings. The other engine severed an elevator cable while a woman was riding the elevator car and followed it down the shaft, landing on top of the car. The woman was rescued but eleven workers from the National Catholic Welfare Conference died along with three people on the plane. Casualties were limited as the accident occurred on a Saturday and few people were inside. The building suffered about a million dollars in damages but was structurally sound.[7]

"Guess we just missed seeing that accident at the Empire State Building," Art wrote from the ship. "It happened soon after we passed it. It must have been an awful thing though."

The *Greely* sailed into the Atlantic, its destination undisclosed. Flo heard rumors the ship was again bound for Le Havre but began to doubt them. "But wish you were. It would be good to see you back again quick-like." Until then, music and letters would bond them. "I get very homesick whenever I hear a band," she wrote. "I just can't wait until you're the conductor and I'm the French horn player again."

The *Greely* had a secret mission on this Atlantic crossing that surely baffled crew and troops who observed unusual operations on deck. Her War Diary[8] notes that in accordance with confidential orders, she dropped one or two experimental bombs overboard on each day beginning July 29 until August 5. Weighing only four pounds, the bombs were incapable of damaging an enemy vessel, and Nazi submarines no longer patrolled the Atlantic. Three weeks after the *Greely* dropped the last little bomb, a Woods

Hole Oceanographic Institution report outlined the purpose of the bombs. Working with Woods Hole Oceanographic Institution, the Navy tested a system for signaling the location of distressed ships, including sailors in lifeboats by exploding four pounds of TNT at a depth of 4,000 feet. The sound waves from the explosive charge travel about a mile a second and could be picked up by hydrophones on ships or near land by a receiving device in deep water if there were no land masses between the explosion and the receiver.[9] A Woods Hole report concluded, "Recommendation is made to utilize a network of monitoring stations to locate planes, ships, and life rafts in distress on the open oceans. Three or more stations receiving a signal could locate the source better than one mile."[10]

In the safety of the Atlantic, officers and crew of the *Greely* had no idea how important such a rescue system would have been in the Pacific, where Japanese submarine I-58 under command of Mochitsura Hashimoto prowled the waters between Guam and the Philippines—where Captain Charles McVay directed the U.S.S. *Indianapolis*, a heavy cruiser with 1,196 men aboard. At the island of Tinian in the Northern Marianas, the *Indianapolis* delivered the components of Little Man, the atomic bomb that soon would be exploded over the Hiroshima, Japan. She then departed for Navy headquarters in Guam, where she was ordered to sail west to Leyte Gulf, Philippines, to meet the battleship U.S.S. *Idaho* to prepare for the invasion of Japan. The *Indianapolis* was halfway to the Philippines when circumstances favored Hashimoto—the timing and location of the sub surfacing, the light from the rising moon, the parting of clouds near the horizon, the huge ship looming before the submarine without destroyer escort. It was the ultimate shot in Hashimoto's four years of service. Two torpedoes[11] sank the ship in about twelve minutes, taking 300 sailors to the bottom. The rest entered dangerous waters with few lifeboats or flotation devices. Many died from shark attacks, injuries, drownings, or hypothermia. Although the Navy picked up distress signals, it did not act on them and did not consider the ship's overdue schedule a sign of disaster. When survivors were spotted in open water five days later, a massive rescue operation was launched, saving 316 of the almost 900 who escaped the ship. In February 1946 McVay was found guilty of negligence for giving the order to cease a zigzag course in clear visibility, a strategy for avoiding torpedoes that Hashimoto testified would not have saved the ship. With clemency, McVay was later promoted to rear admiral. He committed suicide in 1968 and was later exonerated by an act of Congress.[12]

News of the nation's worst naval disaster at sea did not immediately reach the sailors on the *Greely* or the American public. The musicians were as busy as they had ever been with daily chores, oblivious to the enormous tragedy. Art and Tom Stokes worked the canteen at select hours for the troops and officers. Between rehearsals and performances, the bands played rousing numbers on deck to a crowd of troops.

On August 6, 1945, as the *Greely* was steaming east through the Mediterranean about sixty miles north of Algiers, an American B-29 bomber dropped the atomic bomb delivered by the *Indianapolis* over the Japanese city of Hiroshima, killing an estimated 80,000 people. Many more would die from radiation exposure. Three days later, a second B-29 dropped another atomic bomb on Nagasaki, killing an estimated 40,000 people. News reached the *Greely* quickly, and for most of the troops, the bombings heralded the end of the war and would spare them the most dangerous assault of the Pacific war. Still, the fate of the *Indianapolis* was secret.

"*Greely* War Diary," August 15, 1945
Upon announcement by the Commanding officer that war with Japan was officially terminated by reason of unconditional surrender of Japan, official honors were rendered, a twenty-one (21) gun salute was fired, and the day declared a holiday for the ship's crew. Expended twenty-one (21) rounds of 5" 38 ammunition; no casualties.[13]

The Navy finally revealed the sinking of the *Indianapolis*. The devastating news of the enormous loss of life two weeks before the end of the war was overshadowed by the euphoria of peace, as likely it was intended.

Flo's letter two days later expressed the relief and joy of a nation. "Happy day! I never thought I'd see it. Peace is wonderful. We had two days' holiday and the gas rationing is over." With people filling their gasoline tanks, Flo had difficulty finding a station that had fuel. She rounded up her friend Fran and mother, Francine, and Genny, and headed for Alley Pond Park in Queens, where they rented bicycles and rode all over. It was a perfect day except their husbands were missing. "Everything else was wonderful. The weather was nice and the sky was so blue. I even think nature was celebrating yesterday."

Most of Art's letters to Flo regarding the war's end are missing. His "happy letter" to his mother suggested no furtive fears, only hope. "The future of this old world looks very good. I hope they have all learned their lessons and there will be no more wars because with these atomic bombs, it would be no joke."

The war was over as the *Greely* sailed for India with more than 3,000 officers and fresh troops that had little or no patriotic purpose to mitigate the pain of separation from families. If crew members believed that Stedman would waive military protocol in the absence of war, they were wrong. Four crew members were caught gambling. As petty gambling was common on the ship with no indication of an infraction of Navy protocol, it is likely these four were gambling on a larger scale. The commanding officer convened a captain's mast and sentenced each to twelve hours extra duty[14] as the *Greely* made its way up the Hooghly River to Calcutta. After passengers debarked and cargo was offloaded, the ship filled with more than 3,000 officers and troops to begin the voyage home August 30. Of the 3,075 passengers aboard, 252 were black officers and three were civilian black women, an indication of the war's recognition of the contribution of African Americans.[15]

Integration was a hard sell in the early days of the war. Little over a month after the attack on Pearl Harbor, the Navy's General Board mulled the question of recruiting African Americans to serve in roles other than messmen without consensus. Navy Admiral Charles Snyder broached the topic of recruiting African Americans for music programs. "It has been suggested that possibly we might open up the musician branch. The colored race is very musical and they are versed in all forms of rhythm,"[16] he told the board. Receiving no response, he again raised the question at the end of the meeting. "What about the musicians I suggested? Could we open up that field to them?" Navy Captain Harry Asher Badt, director of the Enlisted Personnel Division, quickly torpedoed the idea. But by the spring of 1942, the Navy began accepting African American musicians, who eventually made up more than 100 of the 285 Navy bands during the war.[17] The Coast

Guard was receptive to integration. The Manhattan Beach Training Station was the first military organization to racially integrate with whites and African Americans sharing living and training facilities after accepting 150 African Americans in the spring of 1942.[18] The federal government selected the Coast Guard as the experimental vanguard of military desegregation with the training of officers. In 1943, African American officer candidates began training in the Coast Guard Reserve Officer Training Program. By the end of the year, the first African American officers and enlisted personnel began serving on Coast Guard ships. By the war's end, 5,000 African Americans had served in the Coast Guard, some receiving medals including the Bronze Star Medal, Navy & Marine Corps Medal, Navy Commendation Medal, Silver Lifesaving Medal, and Purple Heart Medal.[19]

There is no indication in the record that African Americans served in the Manhattan Beach bands. Racial barriers fell throughout the military during the war. For example, the Army recruited African Americans who served with distinction during the desperate days of the Battle of the Bulge[20] and in other battles late in the European war.

As the *Greely* sailed south September 2 off the east coast of Ceylon, Japanese envoys Foreign Minister Mamoru Shigemitsu and General Yoshijiro Umezu boarded the U.S.S. *Missouri* to sign the eight-paragraph document of unconditional surrender, accepted and signed by Gen. Douglas MacArthur, commander of the Southwest Pacific Theater and Supreme Commander for the Allied Powers.[21] Facing near gale-force winds, the *Greely* sailed around the southern tip of Ceylon and up the west coast to Colombo Harbor, Ceylon, where she spent a day at mooring. The next day, she sailed into another two days of rough weather, rolling, pitching, and taking spray over the port bow. The *Greely Grenadiers* took a break from morale-building concerts on deck until the seas settled. Amid the joy and excitement of going home without the risk of a torpedo attack, the passengers endured the rough seas with few complaints. They had endured far worse conditions. The *Greely* reached the comfort of the Suez Canal September 12, cruised through the Mediterranean, and crossed the Atlantic Ocean, arriving at New York Harbor September 26, docking at Pier 84.

CHAPTER NINETEEN

Bringing Them Home

Sgt. William F. Baker of San Antonio, Texas, and Sgt. William E. Abel of Philadelphia each debarked the U.S.S. *General A.W. Greely* with a new wife and baby, the beginnings of a generation conceived in war and destined to enjoy a relatively peaceful future in the United States.[1] The two sergeants were among 3,075 troops[2] and civilians who arrived in New York September 26 from the China-Burma-India theater along with bags of mail and cargo. Most who debarked the *Greely* that day were among the longest serving troops in the war, some more than four years away from home, others sick and exhausted from jungle warfare, 350 of them hospitalized before and during the voyage.[3]

The New York Times announced the *Greely*'s anticipated arrival with a warning: "Port officers requested relatives of returning troops not attempt to meet the ships, but to stay at home awaiting messages from men coming in."[4] Expecting families and loved ones of those on the *Greely* to show restraint in waiting at home was nothing less than folly. They had endured separation and anxiety, knowing their loved ones fought on the most dangerous battlefields of the war. They crowded Pier 84 to get a glimpse of the Army heroes of the China-Burma-India theater parading down the gangway to the music of the *Greely Grenadiers*.

"Laden with a rich store of heroism and high adventure in exotic lands, the Army transport *General A.W. Greely* arrived here yesterday, docking at West Forty-Fourth Street with 2,775 men and women of the armed services, the first large group to return since V-J Day from its China, Burma, India war area," *The New York Times* reported.[5] The *New York Mirror*'s headline read, "GI Jungle Heroes Home from India" beneath a photograph of the crowded deck of the *Greely*. Tapestry adorned the side of the ship bearing the insignias of more than a dozen fighting forces and support groups with the caption "First C.B.I. [China, Burma, India] veterans home since 'V-J Day.'"[6]

The U.S. Army engineers who built the Ledo Road debarked after three years of what was regarded as "the toughest job ever given to U.S. Army Engineers in wartime."[7] Over three years beginning December 1942, with help from local labor, the Army engineers carved the road over mountains and through jungles of northern Burma to

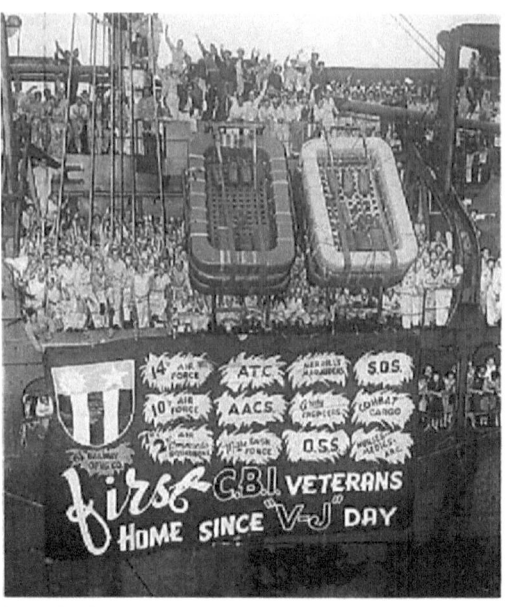

The CBI's longest serving troops debark the *Greely* in New York. The banners represent various military groups including Merrill's Marauders. National Archives

re-establish a supply route to China that had been blocked by the Japanese invasion of Burma.

The Kachin Rangers walked down the gangway to the cheers of the crowd and music from the ship. The Rangers, a unit of Detachment 101, enlisted the Kachin fighters of northern Burma to assist in attacking Japanese installations, a task they were happy to do as bitter enemies of the Japanese. Trained by the United States in guerilla tactics, Kachin Rangers, comprised of U.S. and local fighters, ambushed Japanese patrols, rescued American pilots shot down by the Japanese, and cleared landing strips in the jungle. They proved essential in locating Japanese installations for attack by the Tenth Air Force. The Office of Strategic Services valued their local allies and awarded each native Kachin fighter a Certificate of Military Assistance when the United States troops departed in 1945 aboard the *Greely*. The Tenth Air Force also was deployed in the China, Burma, India theater beginning early 1942.[8] The air operations supported ground forces in battles in northern Burma, where they strafed and bombed Japanese troops, supply depots, communication lines, and artillery installations. Twenty-seven of the Tenth Air Force pilots began their air operations by taking off from the aircraft carrier *Hornet* in Doolittle's 1942 raid on Tokyo, crash-landed in China, and remained in the China-Burma-India theater for the duration of the war.[9]

Pilots from the famed Flying Tigers debarked after as much as four years defending China from air attacks by the Japanese. Flying Tigers were initially volunteers from the Army Air Corps, the Navy, and the Marines who began in 1941 as the American Volunteer Group. With extensive training, these airmen became known as the Flying Tigers and inflicted heavy losses on the Japanese in air attacks over China. The 1st Air Commando Group and the 2nd Army Air Corps Group returned home from the China-Burma-India theater after providing a year of fighter cover, bombardment, and air transportation services for troops operating in Burma, many behind enemy lines.

Merrill's Marauders, code-named "Galahad," were volunteers who answered the call to perform an undisclosed, hazardous mission. Many were less experienced in jungle warfare than organizers had hoped, and some with combat experience suffered from malaria. In all, 3,000 troops trained in long-range penetration and other tactics while exhibiting less discipline than commanders would have liked.[10] Under command of Brig. Gen. Frank D. Merrill, the force began a 1,000-mile trek over mountains and into the Burmese jungle behind Japanese lines to challenge the 18th Japanese Division. Supplied with sporadic airdrops, they hauled heavy equipment over rough jungle terrain to attack Japanese installations and an airfield. Although greatly outnumbered, the Marauders decimated Japanese forces.

Exhausted, the survivors clung to a promise of relief and rest, a promise broken by theater commander General Joseph W. Stilwell, who pushed the Marauders beyond the limits of human endurance, causing bitterness and mistrust of the military command. By the end of May 1944, the Marauders were losing 75 to 100 men a day from malaria, dysentery, and typhus. Merrill himself suffered two heart attacks. When the fighting was over, many Marauders were hospitalized in India with dysentery, malnutrition, malaria, and injuries.[11] When the *Greely* arrived in Calcutta, they were among the troops given priority repatriation. Many were among the 350 hospital patients noted in the *"Greely* War Diary."

Years later, my father reminisced about that voyage, the ship laden with war heroes, some thrilled to be going home, some injured, some damaged by the extremes of jungle warfare and the loss of so many comrades. Some of Merrill's Marauders were so disillusioned by Stilwell's failure to provide them rest and recuperation they felt abandoned. Some made little sense in conversations. Others suffered passively from the diseases and hardships of the past couple years. Their lives had been a fight against overwhelming odds, where the rigors of military protocol were lost in the struggle to survive while fighting against overwhelming odds. Many went about the ship without regard to discipline and acted recklessly under the yoke of military protocol.

Commander Stedman demonstrated empathy for his passengers, understanding the hardship they had endured and the damage they had sustained. Besides, troops expecting repatriation were rioting throughout the world. The war was over and so was military discipline. My father recalled that Stedman ordered the ship's officers and crew to let the Marauders do as they pleased unless they were likely to harm themselves or others. The full story of their accomplishments and hardships had not yet been told. The military was intent on keeping news of their psychological damage under wraps. *Let There Be Light*, John Huston's 1945 documentary of post-war psychological trauma, documented the suffering of returning war veterans with psychiatric conditions treated at an Army hospital in Long Island, New York. Huston, a major in the Army Signal Corps at the time, revealed war damage the military was not prepared to acknowledge. Blocked by an Army directive, the documentary was not released until 1980.

Euphoria over the end of the war was quickly tempered by the realization that homecoming was not forthcoming for many in the military. Art's letter of August 20 reflected the frustration of musicians, sailors, troops, nurses, and home-front families. "Tonight, I am very lonesome, darling, more so than usual, guess it's just the idea that the war is over and still we have to be separated. There's one thing about it, honey, they can't keep me too much longer." In his optimism, Art ignored the fact that he served on a troop transport ship, and hundreds of thousands of troops were waiting to come home.

Over the previous four years, the United States had transported an estimated 7.5 million troops to foreign lands, and it would take more than a few months to bring them home. Troops abroad and families at home had no patience for the slow pace of repatriation. No longer part of a patriotic cause, troops abroad were beset by loneliness, boredom, and existence without purpose. In foreign ports and bases in Europe and the Pacific area, frustrated troops with time on their hands turned to protests. No longer respecting military authority, many engaged in rebellious behavior that officers were powerless to quell. R. Alton Lee, author of "The Army Mutiny of 1946," wrote that the conduct of thousands of troops amounted to mutiny, an offense punishable by death, a sentence that was not carried out.[12] The sheer number of troops was not the only reason for delays in repatriation. The Truman administration, now at the pinnacle of global power, had reason to station troops throughout the world to maintain U.S. dominance.[13]

Republican politicians, eager to erode the Democratic stronghold developed during the war, egged on troops and activists at home to protest the Truman administration for dragging its feet in repatriation. The rallying cry in streets throughout the United States was "Bring the Boys Back Home," with more than 200 groups echoing the cry and pressuring Congressional representatives with pictures of loved ones and baby shoes labeled "Please bring back my daddy."[14]

Eight months before the fall of Germany, the War Department developed a plan for repatriation, establishing priority for homecoming based on a point system that counted length of service, overseas deployment, combat duty, and parenthood. Soldiers with 85 points or more were first in line to go home. Female military personnel needed fewer points to get top priority. Initially, there was a sense of agreement that the rating system was the best among difficult options. The failure of the point system became apparent by the end of the war when troop transport ships lacked the capacity to bring home many with the highest ratings.

In the CBI theater, troops vowed political revenge over the delays, touting the slogan, "No boats, no votes." In one Army mailroom in Tokyo, enlisted men stamped homebound letters with that slogan. Citing Navy statistics, *The New York Times* noted that of the 3.4 million men in the Pacific on September 2, more than 1.1 million had been shipped home by the end of November. That was not good enough. Of the 150 letters a day received by the Stars and Stripes office in Tokyo, more than half were complaints about the slow pace of repatriation. Soldiers complained about the unfairness of the point system, the merchant cargo ships that went home empty, the Army's reluctance to pack soldiers into homebound transports that were packed on their voyage to the CBI theater.[15] A day after the publication of The New York Times's "Pacific Veterans Press for Return," the Army responded that thirty additional ships would be deployed to the Pacific theater to accelerate repatriation.[16]

Stedman was agreeable to relieving as many crew and officers as possible. He surveyed his crew after Japan surrendered to determine who wanted to leave and who had discharge priority. He wrote the Coast Guard commandant, "It is the Commanding Officer's firm conviction this vessel can be operated efficiently with a substantial reduction in officers and enlisted men."[17] Musicians, observing the euphoria of the homebound troops, estimated their own points and calculated when they, too, might be home after the war. Of the more than 400 enlisted crew on the *Greely*, only one indicated an interest in continuing on the ship. Only fifty-three had earned the 43 points that qualified them for priority discharge. Four of them were musicians. The remaining twenty musicians would have to wait. Art had only 37 ½ points. Roger Hartman, with only 31 points, was again stymied in his efforts to abandon ship.[18] The best hope for many of the *Greely Grenadiers* was to be home by Christmas.

As Americans on the home front adjusted to a life of peace, many did not envision a future of revolutionary values that had evolved during the hardships of war. Instead, in victory and peace they returned to old values in a prosperous economy. John W. Jeffries, author of *Wartime America: The World War II Homefront*, wrote, "Visions of peace focused especially on home, family, jobs, and the good life of abundance and security denied for fifteen long years of depression and war. Just as wartime Americans often resisted home-front change, so most of them wanted a postwar future that was essentially a more prosperous and secure version of the past."[19]

The daily anxiety of separation and the fear of losing a loved one in battle withered, as did the summer of 1945. Concerns of danger, life and death, location of loved ones, anxiety over impending battles, and waiting for love letters indicating the sender was alive at the time of post quickly became a distant past. Families reunited after the war adjusted their lifestyles to almost unlimited fuel, food, and other commodities that had been too costly during the Depression or rationed during the war. Troops and civilians planned their futures in a war-free existence.

For many military families, those plans emerged in separate venues, awaiting a post-war reunion. As sailors on a troop transport, Art and fellow musicians would be engaged in naval operations for the next few months biding time to enter a new lifestyle. Flo responded to Art's earlier suggestion of a September rendezvous with a raincheck request. "Sometimes I wonder about that September deal," she wrote. "We had better change it to November anyway." Even the November rendezvous was optimistic. The *Greely* made a third trip to India, with the musicians playing for home-bound troops.

Censorship was over. No more deciphering "Dear Darling letters." "I read in the paper that all censorship is removed, so you can write what you want except that by the time you get this, you won't be writing anymore," Flo wrote in the last surviving war letter on September 4. "I wish I knew where you are right now," she continued. "It's awful just waiting here. Hope you're on the way home when I'm writing this. Goodness, I just can't wait."

As the riots abroad peaked, Art was discharged December 20, in time for Christmas. Flo's summer job as a playground monitor was coming to an end, and she would be "adrift again," contemplating how to finance the couple's future plans. "I have to make lots of money for our bank account. It will be over $1,000 by the time you get home," she wrote. The *Greely* wives continued to meet while their husbands served on the ship. They got together for supper at Flo's apartment, where she settled into a rocking chair that collapsed with a loud crack as she hit the floor. "I could have died," she wrote. "We laughed so hard." News from the home front focused on the mundane. The battery in the car went out again. Her brother, Fred, replaced a blown fuse in her car and got the headlights working. He threatened to fix the gas gauge, too, but did not by the time her letter was mailed. Flo's father, George, was tormenting little Francie again, making faces at her so she would yell. "There's not much to write about anymore as nothing ever happens," she wrote, trying to fill the page.

Art quit smoking but he worried he would get fat as he was eating far more than he ate as a smoker. Romance would soon be a routine of marriage rather than a precious moment salvaged from the wreckage of war. He wrote Flo a provocative post-war letter that was lost and solicited a response in a follow-up letter. "How did you like the ending on last nite's letter—a little exciting sexually anyway. Do you know, cutie, this life is a very sexless existence, but in spite of that I haven't lost my sex for you, honey. Shocking, isn't it? I can just see you blush. That's the way I feel, rather shocking."

It should not have been. It seems many in the nation shared his sexual interests, with war couples' bedroom reunions leading to a generation known as "Baby Boomers." Post-war sexual activity produced a population explosion unlike any other in the nation's history. The two sergeants and their wives who boarded the *Greely* in India were out of the gate early in the reproduction race as were George and Betty Bland and other war couples. Population growth figures show the rest of the war generation was not far behind. In 1946, 3.4 million babies were born, 20 percent more than in 1945. In 1947, another 3.8 million were born, followed by 3.9 million in 1952. Between 1947 and 1952, more than 4 million were born every year,[20] the same period when Art and Flo's four children were born, the first in March of 1947.

While most *Greely Grenadiers* continued to serve in the Coast Guard after Japan's surrender, sea duty was over. In early October, Arthur Schnell and six other musicians were transferred to Ellis Island for land-based assignments until their discharge. The remaining musicians were transferred to the Manhattan Beach Training Station.[21] The *Greely* sailed its last voyages without the music of the *Greely Grenadiers*.

CHAPTER TWENTY

After the War

The U.S.S. *General A.W. Greely* made a final trip to Calcutta, leaving New York December 14 and returning with troops February 2, 1946. She sailed her last war voyage without the *Grenadiers*. The musicians were discharged with the American Area Campaign Medal, the Asiatic Pacific Area Campaign Medal, the European African Middle Eastern Area Campaign Medal, and the Victory Medal. Those medals did not help them get shore duty while the war was on, but they did honor their service for years to come.

Control of the Coast Guard was returned to the Treasury Department in 1946, voiding its naval combat role and reestablishing its traditional role of port security and rescue. The *Greely* was decommissioned in San Francisco and assigned to the War Shipping Administration for use as an Army transport. The Navy took her back in 1950 for use by the Military Sea Transportation Service with a civilian crew, transporting passengers and cargo between Europe and the United States. Following the outbreak of the Korean War, the *Greely* returned to her troop-transport role. A decade later, she was sold and converted to a private container ship. She was scrapped in 1960.

Commander George W. Stedman Jr.

Commander George W. Stedman Jr. continued to serve in the Coast Guard with various assignments on the West and East coasts. In February 1952, Stedman was assigned to command the *Tamaroa*, a cutter with a history and a future of dramatic rescues based in New York. His one-year stint as captain ended decades before the *Tamaroa* earned fame for her rescue of three sailors and four downed Air National Guard crew documented in The Perfect Storm. Her participation in the 1956 collision of the Stockholm and the *Andrea Doria* got barely a mention in the stories and books of that disaster,[1] although she was among the first ships on the scene and escorted the crippled Stockholm into New York harbor.[2] By 1960, we find Commander Stedman head of the Coast Guard port operations in Boston, where he oversaw various rescue operations including the crash of a commercial airliner in Boston Harbor October 24. During the nightlong operation, eleven people were rescued.

George Bland

Among the many musicians who collaborated with my father on the ship, George Bland was his closest friend. Art maintained communication with George after the war and visited him at his home in Weston, West Virginia, along with spouses and Joe Williams, drummer and drum major from the Manhattan Beach band who had not been assigned to the *Greely*. Bland continued his interest in music but changed careers after the war, working for Citizens Bank of Weston for thirty-five years. George and Betty Bland inherited a number of items after the death of George's father, a Texas resident and the son of a famous Texas painter. Bland and his family packed up inherited items and stored them in their attic in Weston.

Since childhood, Jon Buell loved to poke around in his grandparents' attic. It was full of old stuff that seemed more curious than valuable. As many times as Buell had surveyed the attic contents over the years, he had overlooked a painting covered with a tarp. In 2009, as an adult of 32, Buell uncovered a 5-foot by 7-foot painting and observed a mural depicting the Battle of San Jacinto. It was signed by Henry Arthur McArdle, a Texas painter who was Jon's great, great grandfather.

"I knew my great, great grandfather was a famous artist known for his epic murals of Texas battles," he told Jamie Colby for Fox's Strange Inheritances show.[3] "I had no idea I'd end up solving a mystery surrounding one of them." His grandparents doubted the painting had any value although they knew McArdle was the well-known artist. Buell was determined to have the painting evaluated.

Henry McArdle immigrated to the United States from Ireland at the age of 14. During the Civil War, he served as a draftsman and map maker.

After painting portraits of Confederate soldiers, McArdle looked to earlier times for subjects and focused on the Battle of San Jacinto. The battle took place after the defeat of Texas forces at the Alamo. Sam Houston led the charge as about 900 Texans made a surprise attack on Gen. Antonio Lopez de Santa Anna's forces at the San Jacinto River. Their battle cry was "Remember the Alamo." With the defeat of Santa Anna in 1836, Texas gained independence from Mexico. Forty years later, Henry McArdle made it his mission to depict the battle with a mural 14 feet wide and 8 feet high, completed in 1898 after a decade of work. It was regarded as a masterpiece and was displayed in the Texas Capitol Senate chamber along with five other pieces, confirming Henry McArdle as legend in Texas history. But the Senate never paid for his paintings, which are still displayed there.[4]

An art patron named J.T. DeShields commissioned a smaller painting of the "Battle of San Jacinto" in 1901. As the $400 fee was never paid, McArdle kept the painting. From there, its whereabouts was a mystery.[5] The book *Painting Texas History to 1900* contains the only known reference to the painting. Author and historian Sam DeShong Ratcliffe concluded, "location unknown."[6] It remained in the Blands' attic until Jon Buell uncovered it and had it evaluated. George Bland lived long enough to appreciate the historical value of the painting but not long enough to realize the economic value. He died November 15, 2009, at the age of 92. A year later, the painting sold at auction for $334,600. The buyer loaned the painting to the Public Policy Building in Austin for public display. There it became the only surviving relic of the McArdle attic stash, as fire destroyed the home in 2015, after which Betty Bland moved to Port Orange, Florida.

Eli Bublick

Among the most talented of the *Greely Grenadiers* was trombonist Eli Bublick, a mentor to my father and an entertainer with or without his trombone. He immigrated to the United States from Poland at the age of six weeks. Before joining the Coast Guard band, Bublick played in various big bands during the 1940s. After the war, he worked in clothing manufacturing and joined his wife in their retail store in Passaic, N.J. He retired to Florida

1984 and worked as field representative for the U.S. Census Bureau. He was a lifetime member of the Musicians Union.

Shelly Manne

Shelly Manne is regarded as one of the greatest jazz drummers in history. His music contributed to thousands of records and Hollywood movies. While a drummer with the Coast Guard in Manhattan Beach, Manne mingled with top musicians at New York jazz clubs. After the war, Manne went on the road with Stan Kenton's band, cut records and played with top musicians, including Ella Fitzgerald. In 1952, Manne and wife Florence "Flip" Manne moved to the West Coast, where he became a leader in the jazz movement. He was a favorite choice for Hollywood movie scores because he could adapt his style to fit the circumstances. Manne moved beyond performing and recording, opening a Los Angeles club in the summer of 1960 that became a magnet for great jazz performers. He died September 26, 1984, just after being honored by Los Angeles Mayor Tom Bradley and the Hollywood Arts Council, who declared the date "Shelly Manne Day."

Clare Grundman

The *Greely Grenadiers* admired their musical chief Clare Grundman, who oversaw the Coast Guard bands during the war. Grundman was one of the most influential composers of band and wind ensemble repertoire, with works dating from the 1940s. I found no indication that Art had met him, but his leadership role is mentioned throughout the letters. Before the war, Grundman performed in orchestras on ships crossing the Atlantic, including the French ship S.S. *Normandie*, before she turned turtle in the Manhattan berth. After the war, he returned to civilian music, focusing his career on musical composition for radio, television, motion picture, ballet, and Broadway performance. His musical talents were recognized with numerous awards, by many music associations including the American Bandmasters Association, where he was inducted in 1985.

William Schallen

The *Greely Grenadier*'s musical director returned to New York after the war where he pursued his musical career with big-name bands such as the Tommy Dorsey Orchestra. He also worked as a studio musician, performed in television bands for *Dave Garroway Show*, the *Steve Allen Show*, the *Merv Griffin Show*, the *Johnny Carson Show*, and others. His musical performances were featured in Broadway shows and motion pictures productions.

Roger Hartman

Aboard the ship, my father enjoyed the company of musician Roger Hartman, the nervous fellow who feigned kidney problems and kept a stash of Hostess Twinkies to fortify him if he ended up in a lifeboat. During idle times on the ship, they sat together and talked, my father discussing music and his wife, Roger plotting course to the nearest friendly island. Roger visited us in Van Hornesville many years after the war. I was quite young and do not remember much about him or the conversations. He stayed the night, but the next morning when we got up, he was gone without a goodbye or a note. My parents were baffled, and we didn't hear from Roger again.

Arnold Broido

Arnold Broido, who was instrumental in Art's entry in Coast Guard performing bands, was a piano prodigy who intended to be a music teacher after the war. Unable to find a job, he worked in music publishing, ultimately serving as president of Theodore Presser Co., publisher of symphonic and concert music. He championed the rights of musicians as photocopy machines and the Internet threatened composers' revenues.

Belmont Ketchel

Belmont Ketchel continued to pursue his love of dance music, making a career of performing in the clubs of New York after leaving the service.

A 1950 Census report lists him working as a nightclub musician. He and Lillian Ketchel divorced in 1951.

Arthur and Florence Schnell

Through war letters, Art and Flo expressed their mutual frustration with crowds of New York, long lines on a crowded ship, uncomfortable train rides to work and back, desiring to live in the country after the war. They agreed to buy a house, in early references, a simple cottage, with Flo building up the bank account. After Art's discharge, they headed north to Upstate New York. At first, Art considered getting a master's degree but then accepted a music teaching position at Van Hornesville Central School, now Owen D. Young Central School, where he had worked a year before enlisting. It was small by all measures, with a K-12 enrollment normally between 300 and 350 students. He taught there for thirty-two years, his bands giving annual concerts, marching in Fourth of July parades, and playing for most community events except football games, as he did not like team sports. His bands competed each year in Allstate competition and earned some top honors. Art never lost his love of the big bands, continuing to idolize Tommy Dorsey and Glenn Miller, and playing in local dance bands for proms, weddings, and other events. Many nights I heard him practicing dance tunes in the basement and had to admit to myself that I loved his melodies. But he did not love mine. I listened to Bob Dylan and other folk singers of the 60s that he considered devoid of musical talent and did not hesitate to tell me. Once while Bob Dylan was playing, he entered my room and said, "Jesus, that guy sounds like he's sitting on the can with a bad case of constipation. Give him an enema and get it over with." Later, as I listened to his music drift up from the basement, I heard what I considered the most beautiful melody. Years later I learned it was "Stardust" by Hoagy Carmichael, regarded as one the best popular and most recorded tunes in musical history. I could never tell him, even when he was diagnosed with terminal cancer, that the music was beautiful, certainly more beautiful than what I was listening to. He was so competitive that I would not have heard the end of it, that he was right all along, that my music was trash. As musical tastes changed, I heard him practicing *Beatles* music for an upcoming prom. I scored a silent victory.

Flo taught vocal music to elementary pupils and directed the school chorus. She also directed the choir at the local Methodist Church and gave private music lessons. The couple performed in a local symphony based in Little Falls, New York, filling the trombone section. After my father's death in 1996, my mother continued to play trombone in a community band in Cooperstown.

Over the course of three war years and more than 100 letters, the couple settled on a family of three. They exceeded the three-children agreement by one, the fourth, William Schnell, assumed to be an accident but a fortuitous one as it turned out. The oldest, Arthur Schnell Jr., severed ties with his parents after college and remained estranged during the rest of their lives. So, the youngest redeemed the prophecy of the letters, becoming the third of three after serving for some twenty years as an alternate. In 1952, they built their home, adding a couple bedrooms as the kids came along.

Sailboats and sailing dominated much of our lives growing up. The first sailboat was a 12-foot catboat my parents kept at the north end of Otsego Lake. I was small and the boat seemed large. On annual sailing trips to Cooperstown at the south end, I would crawl under the tiny foredeck and imagine I was on a big sailboat crossing an ocean, heading for exotic places as my father had done during the war. Years later, Art founded the Otsego Sailing Club and served as the first commodore. The club was originally intended to promote sailboat racing, and my father organized a fleet of Rhodes Bantams, a 14-foot boat that was popular throughout the Finger Lakes, Lake Erie, and other locations. He contributed to the fleet newsletter. While the Otsego Sailing Club hosted races, we often traveled with sailboat in tow to other lakes for regional and national competitions in locations as near as the Finger Lakes as far south as Biloxi, west to Missouri.

Thus, three of the four endeavors that defined my parents' post-war years were established by war letters that traveled to Upstate New York to their permanent repository in the Utica bacon box. During boot camp and in the urban rat race of New York City, Art came to appreciate Van Hornesville, the pastoral setting, the natural amenities of Upstate New York. "I'll really appreciate a good many things of life when I get out of here," he wrote Flo during the war. And he did. He and Flo became naturalists and environmental activists who maintained a love of Upstate New York, even after becoming snowbirds wintering in Vero Beach, Florida. They backpacked parts of the Appalachian Trail, canoed the Suwannee River, and engaged in environmental activism in Florida and New York. Although in retirement they traveled through Europe, parts of South America, Utah and Alaska, they never did make that complete cross-country trip they wrote about in 1945.

As the war receded from their lives, the stories too became infrequent until I reached college age. Then there was no more reminiscing of war. I don't believe the war disappeared from their personas as it did from their conversations. Their marriage, like those of Frank, Eli, Shelly, George, and others were lifelong, and I believe the war, the separation, the struggle to be a couple through letters laid the foundation, inspiring them to pursue their passions and their love.

Notes

Preface

[1] "Hollywood and the Office of War Information, 1942-1945," *Backlots*, February 19, 2020. https://backlots.net/2020/02/19/hollywood-and-the-office-of-war-information-1942-1945.

[2] Christine Miller, "Playing a Role in the War: Music in World War II," Concordia Memory Project of Concordia College, n.d. https://concordiamemoryproject.concordiacollegearchives.org/exhibits/show/sartyessays/christinemiller.

[3] Compiled from U.S. Coast Guard and the National Coast Guard Museum sources. For detailed information about the history and services of the Coast Guard, visit https://www.history.uscg.mil and https://www.history.uscg.mil/Research/THE-LONG-BLUE-LINE, and https://nationalcoastguardmuseum.org.

Chapter 1

[1] "U.S.S. *Greely*," undated ship's newsletter published onboard during the first voyage.

[2] Kimberly Guise, "Mail Call: Letters from the Archives." The National WWII Museum, March 7, 2018. https://www.nationalww2museum.org/war/articles/mail-call-letters-archives. Para 8.

Chapter 2

[1] Marshall McLuhan, *Understanding Media: The Extensions of Man* (The MIT Press, 1964).

[2] Guise, "Mail Call: Letters from the Archives," para. 8.

[3] Guise, "Mail Call: Letters from the Archives," para. 4.

Chapter 3

[1] William H. Young and Nancy K. Young, *Music of the World War II Era* (Greenwood Press, 2008), 24.

[2] Young and Young, Music.

[3] Kathleen E.R Smith, *God Bless America: Tin Pan Alley Goes to War* (University Press of Kentucky, 2003).

[4] Young and Young, *Music*.

[5] Smith, *God Bless America*.

[6] Young and Young, *Music*.

[7] Young and Young, *Music*.

[8] Young and Young, *Music*.

[9] Joseph Connor, "World War II Soldiers Loved to Sing—Provided They Got to Sing Their Way," Historynet, n.d.), para. 2. https://www.historynet.com/army-songs-in-world-war-2. The author lists numerous parodies of World War II songs. For a collection and analyses of similar war songs in various wars throughout the world, see Dark Laughter: War in Song and Popular Culture by Les Cleveland, Praeger Publishers, 1994.

[10] Young and Young, *Music*, 21.

[11] Young and Young, *Music*, 25.

[12] Pamela M. Potter, Christina Baade, and Roberta Marvin Montemorra, *Music in World War II*, (Indiana University Press, 2020).

[13] McKenzie Kellar, "Songs for America: Patriotism through Music." *Music Politics* (University of Alabama College of Arts & Sciences, 2017). https://www.musicpolitics.as.ua.edu/projects/primary-source-projects/songs-for-america-patriotism-through-music

[14] "Music is Strong Builder of Morale, Station's Touring Unit Discovers," *Harpoon*, Manhattan Beach Training Station, July 1, 1943, Volume II, no. 9.

[15] Allan M. Winkler (2000) *Home Front: America During World War II*, Harlan Davidson Inc., 14.
[16] Winkler, *Home Front*, 24.
[17] Young and Young, *Music*, 131.

Chapter 4
[1] Mrs. Frank E. Schoen to Admiral Ernest J. King, commander of the U.S. Fleet, June 27, 1942, National Archives II, College Park, Md.
[2] Commander J.R. Hinnant to Mrs. Frank E. Schoen, Washington, D.C., July 15, 1945, National Archives II, College Park, Md.
[3] Manhattan Beach Commanding Officer G.U. Stewart to Commandant, Manhattan Beach, N.Y., September 29, 1945, National Archives II, College Park, Md.
[4] "Offer of Loan of Musical Instrument," Undated, unsigned Coast Guard contract, National Archives II, College Park, Md.
[5] "Music is a Strong Builder of Morale, Station's Touring Unit Discovers," *Harpoon*, July 1, 1943, Vol. II, No. 9.
[6] Ellis Reed-Hill to Edward Lloyd, Coast Guard Public Relations Office, January 8, 1942, National Archives I, Washington, D.C.
[7] William Walton, "The Spencer gets her Sub," *Harpoon*, July 1, 1943, Vol. II, No. 9.
[8] "Beach Patrol XVII," The Coast Guard at War, Declassified document from Historical Section of the Public Relations Division of the U.S. Coast Guard, May 15, 1945.
[9] "Beach Patrol XVII."
[10] "The Legacy of Ithaca College Bands," Ithaca College, n.d. https://www.ithaca.edu/academics/school-music/ensembles/band/history-bands-ic.
[11] "Sixth War Loan," New York Times, December 19, 1944. 20.
[12] Winkler, *Home Front*, 31.
[13] U.S. Coast Guard Official Dispatch. January 3, 1945, January 10, 1945, January 23, 1945, March 3, 1945. A memo of February 17, 1945, mandates that "all men upon transfer be fully prepared for duty afloat or outside the continental limits." National Archives I, Washington, D.C.

Chapter 5
[1] U.S. Coast Guard Official Dispatch, March 3, 1945, National Archives II, College Park, Md.
[2] Winkler, *Home Front*, 48.
[3] "An Iconic Chicago Landmark Comes to Life," Chicago Union Station, n.d. https:/chicagounionstation.com/about/past.

Chapter 6
[1] Winkler, *Home Front*, 38-44.

Chapter 7
[1] "World War II in San Francisco, California," *Legends of America*, n.d. https://www.legendsofamerica.com/san-francisco-world-war-ii.
[2] Roger Lotchin, "Mobilization for the Duration: The Bay Area in the Good War," National Park Service, n.d. https://www.nps.gov/articles/000/mobilization-for-the-duration-the-bay-area-in-the-good-war.htm?utm_source=article&utm_medium=website&utm_campaign=experience_more&utm_content=small.
[3] Lotchin, "Mobilization."
[4] Winkler, *Home Front*, 16.
[5] "World War II Shipbuilding in the San Francisco Bay Area," National Park Service, n.d. https://www.nps.gov/articles/000/world-war-ii-shipbuilding-in-the-san-francisco-bay-area.htm

6 *"General A. W. Greely* (AP-141)," Naval History and Heritage Command, July 10, 2015. https://www.history.navy.mil/content/history/nhhc/research/histories/ship-histories/danfs/g/general-a-w-Greely-ap-141.html

Chapter 8
1 "BB's Stage Door Canteen at The National WWII Museum," The National WWII Museum, New Orleans, n.d. https://www.nationalww2museum.org/events-programs/bbs-stage-door-canteen
2 Undated newspaper article.

Chapter 9
1 "From Luxury Liner to Troop Transport: Story of a Skipper," *The Salt*, March 21, 1945, Vol. 1, No. 1.
2 "War Damage to the U.S.S. *Etamin*, Preliminary Report of," U.S. Navy, May 1, 1944. https://catalog.archives.gov/id/134367013?objectPage=2
3 "War Damage U.S.S. *Etamin*, Final Report of," U.S. Navy, June 1, 1944. https://catalog.archives.gov/id/78477663?objectPage=3
4 "U.S.S. *Etamin*, AK-93," United States Coast Guard, January 4, 2021. https://www.history.uscg.mil/Browse-by-Topic/Assets/Water/All/Other-Vessels-Non-CG/Article/2461253/uss-etamin-ak-93/.
5 "From Luxury Liner to Troop Transport," *The Salt*, March 21, 1945, Vol. 1, No. 1, 3.
6 "War Damage to the U.S.S. *Etamin*," May 1, 1944. https://catalog.archives.gov/id/134367013?objectPage=2. "War Damage U.S.S. *Etamin*," June 1, 1944. The report refers to the evacuation, not the fatal prior attack.
7 "Editorial," *The Salt*, March 21, 1945, Vol. 1, No. 1, 4.
8 "U.S.S. *Greely*," 5.
9 "U.S.S. *Greely*," 2.
10 "*Greely* First Ship to have its own Band Aboard from the Start," *The Salt*, March 21, 1945, Vol. 1, No. 1, 6.
11 "Vt. National Guard band gets war recognition," Boston.com., January 16, 2012. http://archive.boston.com/news/local/vermont/articles/2012/01/16/vt_national_guard_band_gets_war_recognition/#:~:text=The%2040th%20Army%20band%20shipped%20out%20to%20the,to%20U.S.%20forces%20on%20the%20Island%20of%20Guadalcanal

Chapter 10
1 After the surrender of Japan, only twenty-two musicians were listed as serving on the *Greely*. George W. Stedman Jr., Memo to Commandant, September 10, 1945. National Archives II, College Park, Md.
2 "Tars and Spars, the Coast Guard Show," Playbill, Artcraft Litho and Printing company, New York, n.d. https://www.history.uscg.mil/Portals/1/images/faq/tars_and_spars.pdf?ver=2017-07-11-153815-170
3 "U.S.S. *General A.W. Greely* Deck Log," April 9, 18, 1945. National Archives I, Washington, D.C.

Chapter 11
1 R.S. Tewksbury, "Plan of the Day," Memo to *Greely* crew, April 7, 1945.
2 "What causes seasickness?" National Ocean Service, National Atmospheric and Oceanic Administration, n.d. https://oceanservice.noaa.gov/facts/seasickness.html
3 "Charles Darwin on Discovering Sea Sickness," *The Guardian*, September 26, 2016. https://www.theguardian.com/science/2016/sep/26/charles-darwin-discovering-seasickness-weatherwatch
4 Christopher C. Shaw, "On the Dynamics of motion sickness on a seaway," *The Scientific Monthly*, February 1954, Vol. 78, No. 2, 110-116. https://www.jstor.org/stable/21192
5 Stephen R. Bown, "Scurvy: How a Surgeon, a Mariner, and a Gentleman Solved the Greatest Medical

Mystery of the Age of Sail," Captain Cook Society, 2003. https://www.captaincooksociety.com

[6] Kathleen Suits-Smith, "He feels competition of large part of life," *Evening Telegram*, October 22, 1966, 5. The newspaper profiled both Arthur Schnell and Florence Schnell for their contribution to music.

[7] Arthur Schnell to Florence Schnell, April 12, 1945. Art referred to the Red Cross workers as "men," despite earlier publications that referred to them as "woman."

[8] "U.S.S. *General A.W. Greely* Deck Log," April 13, 1945, National Archives I. As crew were paid in cash, a large amount was needed to complete the voyages.

[9] "U.S.S. *General A.W. Greely* Deck Log," March 18, 1945, the date 15 sailors were sentenced during a captain's mast, most given extra work.

[10] Statement by the President on the Death of Ernie Pyle. Harry S. Truman Library, Museum (National Archives, April 18, 1945). https://www.trumanlibrary.gov/library/public-papers/6/statement-president-death-ernie-pyle

[11] "U.S.S. *Greely*," Undated ship's newsletter, 7.

Chapter 12

[1] "Shelly Manne music," cartoon published in *The Salt*, n.d.

[2] "Troops and Crew Rate the *Greely Grenadiers* '4.0,'" *The Salt*, August 23, 1945.

[3] Italian dictator Benito Mussolini was executed 28 April 1945 while he tried to escape with his mistress through the Swiss border. Captured by local partisans, Mussolini and Claretta Petacci were executed the following afternoon, two days before Adolf Hitler's death. "Benito Mussolini Executed," *History*, A&E Television Networks, November 24, 2009. https://www.history.com/this-day-in-history/benito-mussolini-executed.

[4] Hitler was not shot in the streets of Berlin. In his bunker, he poisoned his wife and dog and shot himself. "Adolf Hitler commits suicide in his underground bunker," History, A&E Television Networks, November 24, 2009. https://www.history.com/this-day-in-history/adolf-hitler-commits-suicide.

Chapter 13

[1] Jane Mersky Leder, *Thanks for the Memories. Love, Sex, and World War II* (Praeger Publishers, 2006), 47.

[2] Historians differ as to the appropriate label "revolution" as the sexual standards that emerged during the war were situational and did not endure.

[3] Mersky Leder, *Thanks for the Memories*, xiii.

[4] Barbara Abel, "Hail Hostess!" United Service Organizations Program Service Publication #522, n.d. 9.

[5] C.P. Trussell, "Deferring of Married Men in Draft Is Written into Allowances Bill," *The New York Times*, June 13, 1942, 1.

[6] Mersky Leder, *Thanks for the Memories*, 18.

[7] Jon Zobenica, "The Greatest Sexual Revolution--How World War II Prefigured the '60s," *The American Scholar*, December 2, 2019. https://theamericanscholar.org/the-greatest-sexual-revolution/.

[8] John McPartland, "Our changing Sexual Code: Endless Variety in Stories and Pictures," *Coronet Magazine*, August 1955. Found in http://www.oldmagazinearticles.com/article-summary/sex_during_ww2#.YtbHfYTMKUm

[9] "Jail for the Faithless," *Newsweek*, August 13, 1945. http://www.oldmagazinearticles.com/ww2-divorces-pdf, 26.

[10] "Jail for the Faithless," 1945, 26.

[11] 100 Years of Marriage and Divorce Statistics in the United States, 1867-1967, December 1973, Human Resources Administration, National Center for Health Statistics, Rockville, Md. https://www.cdc.gov/nchs/data/series/sr_21/sr21_024.pdf.

[12] Jon Zobenica, "The Greatest Sexual Revolution--How World War II Prefigured the '60s."

[13] "Out of Wedlock Births," *Pathfinder*, Dec. 11, 1944, found in http://www.oldmagazinearticles.com/

WW2-illiegitimate-births-pdf, 14.

[14] Tom Barnes, "Personality Studies Show the Difference Between People Who Play Music and Everyone Else," BDG Media, August 15, 2015. https://www.mic.com/articles/124032/personality-studies-show-the-difference-between-people-who-play-music-and-everyone-else/.

[15] "Troops and Crew Rate Grenadiers '4.0,'" *The Salt*, August 23, 1945.

[16] *A Guidebook on Calcutta, Agra, Karachi and Bombay*, 93.

[17] "U.S.S. *General A.W. Greely* Deck Log," August 1945, National Archives I. Washington, D.C.

[18] McPartland, "Our changing Sexual Code," 146. Found in http://www.oldmagazinearticles.com/article-summary/sex_during_ww2#.YtbHfYTMKUm.

Chapter 14

[1] Greg Lange, "President Truman announces V-E Day, victory-in-Europe, on May 8, 1945." History Link, 1999. https://www.historylink.org/File/1279.

[2] "Capt. Stedman's Statement as the *Greely* Ends Maiden Trip," *The Salt*, May 20, 1945, Vol. 1, No. 32, 1.

[3] Army personnel Anthony J. Cerasani, Hubert H. Beaird, Nicholas J. Streshka, Coleman B. Miles Jr., and Alvin H. Karfchner were transferred from a jail in Calcutta to the brig of the Greely. Their offenses were not listed in the "*Greely* War Diary" (May 24, 1945). https://catalog.archives.gov/id/140057351.

Chapter 15

[1] Theodore F. Koop, *Weapon of Silence* (University of Chicago Press, 1946), 59.

[2] Koop, *Weapon of Silence*, 60.

[3] Koop, *Weapon of Silence*, 42.

[4] Koop, *Weapon of Silence*, 60.

[5] Koop, *Weapon of Silence*, 60.

[6] Louis Fiset, "Return to Sender: U.S. Censorship of Enemy Alien Mail in World War II," *Prologue Magazine*, Spring 2001, Vol. 33, No. 1. https://www.archives.gov/publications/prologue/2001/spring/mail-censorship-in-world-war-two-1/, para. 98.

[7] Koop, *Weapon of Silence*.

[8] Executive Order 8985 Establishing the Office of Censorship, December 19, 1941. https://www.presidency.ucsb.edu/documents/executive-order-8985-establishing-the-office-censorship

[9] Executive Order 8985. https://www.presidency.ucsb.edu/documents/executive-order-8985-establishing-the-office-censorship.

[10] Koop, *Weapon of Silence*, 16.

[11] Fiset, "Return to Sender," para 10.

[12] Koop, *Weapon of Silence*, 70.

[13] Koop, *Weapon of Silence*, 70.

[14] Winkler, *Home Front*, 32.

[15] David W. Dunlap, "Photo That Was Hard to Get Published, but Even Harder to Get," *The New York Times*, March 28, 2013). https://archive.nytimes.com/lens.blogs.nytimes.com/2013/03/28/a-photo-that-was-hard-to-get-published-but-even-harder-to-get

[16] "Loose Lips do Sink Ships," *World War II in the Pacific*, October 12, 2001. http://www.ww2pacific.com/congmay.html

[17] "Ninth Air Service Chief Demoted, Sent Home, for Talk About D-Day; Maj. Gen. H.J.F. Miller Named in Washington as 'Broken' by Eisenhower for Remark at London Cocktail Party," *The New York Times*, June 8, 1944, 4. https://www.nytimes.com/1944/06/08/archives/ninth-air-service-chief-demoted-sent-home-for-talk-about-dday-maj.html.

Chapter 16

[1] Judy Litoff Barrett and David C. Smith, "'Will he get my letter?' Popular Portrayals of Mail and Morale During World War II," *Journal of Popular Culture*, Spring 1990, 32:4, 23.

[2] Barrett and Smith, "'Will he get my letter?'" 26. Quoting from a radio broadcast.

[3] *Annual Report of the Postmaster General, for Fiscal Year Ended, June 30, 1942*, Government Printing Office, Washington D.C., 1942, 3.

[4] Guise, "Mail Call."

[5] Barrett and Smith, "'Will he get my letter?'" 23.

[6] "New System for Overseas Mail," *Sparkles* newsletter, November 1, 1944, Vol. 1 No. 29. https://media.defense.gov/2022/Feb/25/2002945223/-1/-1/0/1944_Nov_1_Spars_Sparkles_Newsletter.pdf. The newsletter was published by SPARS, the Coast Guard Women's Auxiliary.

[7] "New System for Overseas Mail."

[8] Guise, "Mail Call: Letters from the Archives," para. 6.

[9] Richards and Banigan, *How to Abandon Ship*, 1943, 84.

[10] "The Sullivan Brothers: Transcripts of Service," Naval History and Heritage Command, November 7, 2017. https://www.history.navy.mil/content/history/nhhc/browse-by-topic/disasters-and-phenomena/the-sullivan-brothers-and-the-assignment-of-family-members0.html.

[11] "How Five Brothers Perished at Sea and Inspired an American Legend," The Sullivan Brothers, July 6, 2020. https://thesullivanbrothers.com/how-five-brothers-perished-at-sea-and-inspired-an-american-legend.

Chapter 17

[1] *Greely* War Diary, June 5 and June 7, 1945.

[2] "U.S.S. *General A.W. Greely* Deck Log," June 8, 1945.

[3] "Chinese Make Top Score in Duplicate Tournament," *The Salt*. Vol 2, No. 21, June 15, 1945.

[4] "Initiate Greelytes into 'Sons of Magellan' Round the World Order," *The Salt*. Vol. 8, No. 27, July 21, 1945.

[5] "Glorious, Illustrious Sons of Magellan," Certificate presented to Arthur Schnell for crossing the equator April 3, 1945. Dated June 20, 1945.

Chapter 18

[1] "Le Havre: World Heritage Site: The Bombing of 1944," UNESCO, n.d. http://unesco.lehavre.fr/en/understand/the-bombings-of-1944.

[2] "*Greely* War Diary," July 1945. https://catalog.archives.gov/id/83528214/.

[3] James Hinton, "The Sinking of the S.S. *Normandie* at NYC's Pier 88," *New York Almanack*, September 23, 2014. https://www.newyorkalmanack.com/2014/09/the-sinking-of-the-s-s-normandie-at-nycs-pier-88

[4] "U.S.S. *General A.W. Greely* Deck Log," July 30, 1945.

[5] A Captain's Mast is a formal hearing to determine the outcome of minor infractions in a non-judiciary hearing.

[6] "*Greely* War Diary," U.S. Navy. July 1945. https://catalog.archives.gov/id/83528214.

[7] "Plane crashes into Empire State Building," History, A&E Television Networks, November 13, 2009. https://www.history.com/this-day-in-history/plane-crashes-into-empire-state-building/.

[8] "*Greely* War Diary," August 1945.

[9] W.M. Ewing, "Sound transmission in sea water," Woods Hole Oceanographic Institution, 1941. https://darchive.mblwhoilibrary.org/server/api/core/bitstreams/2ed8bdd5-b577-59b4-9e42-76f7449ca363/content.

[10] W.M. Ewing and J.L. Worzel, "Long range sound transmission: interim report no. 1, March 1, 1944 - January 20, 1945," Woods Hole Oceanographic Institution, 1945. https://darchive.mblwhoilibrary.

org/entities/publication/976cacef-a9ef-5db4-b3b4-e15232e4a267

[11] Hashimoto maintained during the trial that three torpedoes hit the *Indianapolis*.

[12] Richard Newcomb, *Abandon Ship!: The Saga of the U.S.S. Indianapolis*, the Navy's Greatest Sea Disaster (Henry Holt and Company, 1958).

[13] "*Greely* War Diary," August 1945. National Archives I. https://catalog.archives.gov/id/83528214/.

[14] "U.S.S. *General A.W. Greely* Deck Log," August 22, 1945. The Navy policy on gambling is unclear and largely the responsibility of the commanding officer. As petty gambling was regularly practiced without reprimand, it is likely the four were engaged in a larger gambling operation—Dr. William H. Thiesen, Atlantic-area historian for the U.S. Coast Guard in an email August 22, 2023.

[15] "*Greely* War Diary," August 1945.

[16] General Board meeting transcript. (U.S. Navy, January 23, 1942). https://media.defense.gov/2021/Nov/30/2002900538/-1/-1/0/1942-Gb_Integration_transcript.pdf.

[17] Alex Albright, "African-American Navy Bands of World War II," Alexalbright.works, July 2022. https://alexalbright.works/research/music/african-american-navy-bands-of-world-war-ii.

[18] Edwin Nieves and Sean Kinane, "The Long Blue Line: WWII recruit training at Manhattan Beach and desegregating the U.S. military," United States Coast. Guard, February 10, 2023. https://www.mycg.uscg.mil/News/Article/3259454/the-long-blue-line-wwii-recruit-training-at-manhattan-beach-and-desegregating-t.

[19] William H. Thiesen, "First to serve, first to fight and first to sacrifice—African Americans in the U.S. Coast Guard." MyCG (February 2, 2024). https://www.mycg.uscg.mil/News/Article/3664598

[20] John W. Jeffries, Wartime America: The World War II *Home Front* (Ivan R. Dee Inc., 1996), 111.

[21] "Instrument of Surrender," Records of the U.S. Joint Chiefs of Staff; Record Group 218, September 2, 1945. National Archives. https://catalog.archives.gov/id/1752336.

Chapter 19

[1] "Far East Troops Land with Small Supercargos," *Daily News*, September 27, 1945, 26.

[2] The numbers of returning troops were not consistently reported by newspaper or the *Greely*'s War Diary. The "*Greely*'s War Diary" of August 29 lists embarking 3,075 people including civilians. The War Diary of September 26 refers to debarking in New York 398 officers, 2,337 troops, five Navy sailors, 251 civilian men, 76 civilian women—total 3,067, eight fewer than were listed as embarking in Calcutta. The eight-person discrepancy is not explained. *The New York Times* lists 2,775 troops debarked, apparently referring only to military personnel, not civilians. The corresponding War Diary lists 327 civilians among the passengers debarked, which would bring the total passengers to 3,102 if that was how the newspaper intended to report the numbers. *The New York Mirror* lists the number debarked as 3,075, consistent with the number embarked in Calcutta but not consistent with the War Diary numbers debarked in New York.

[3] "*Greely* War Diary," September 26, 1945, National Archives I. https://catalog.archives.govid/83528214

[4] *New York Times*, September 27, 1945, 12.

[5] *New York Times*, September 27, 1945, 12.

[6] *New York Mirror*, September 27, 1945. The subhead of the *New York Mirror* reads "Gen. Healey lands 3,075 after 28-Day Voyage," an apparent error, as the reference is to the *Greely*. The U.S.S. *Healy* was a destroyer that acted as harbor control vessel at Tokyo Bay until the formal surrender was completed. She left Japan September 5 with troops for the United States, docking at San Diego December 21, 1945. Then she sailed through the Panama Canal to New York, arriving on January 17, 1946. *Healy* (DD-672), Naval History and Heritage Command, March 31, 2016. https://www.history.navy.mil/research/histories/biographies-list.html.

[7] "The Ledo Road," CBI-Theater (2022). http://www.cbi-theater.com/ledoroad/Ledo_Main.html/. The quotation is from General Lewis A. Pick, commander of the engineers.

[8] James Musser, Tenth Air Force (AFRC), January 13, 2019. https://www.afhra.af.mil/About-Us/Fact-Sheets/Display/Article/433512/tenth-air-force-afrc

[9] Tenth Air Force Overview, 10 March 1942 - December 1945, US Army Air Forces in China Burma India, n.d. https://www.aafincbi.com/CBI_10af_home.html.

[10] David W. Hogan, "U.S. Army Special Operations." During the recruiting, one officer noted, "We've got the misfits of half the divisions in the country." (13) The author cites an example of unruly behavior: "From the beginning, the unit was hard to handle; when it moved by rail from Deogarh to the Ledo, for example, one officer found his men shooting out the windows at Indians as if they were riding through the Wild West in the 1870s." (113).

[11] Hogan, "U.S. Army Special Operations."

[12] R. Alton Lee, "The Army Mutiny of 1946," *The Journal of American History*, Oxford Academic, November 30, 1966, Vol. 53, Issue: 3, 555-571.

[13] Sam Perkins, "Why World War II Soldiers Mutinied after V-J Day," *History*, 2022. https://www.history.com/news/world-war-ii-soldiers-mutiny-v-j-day.

[14] Perkins, "Why World War II Soldiers Mutinied."

[15] Perkins, "Why World War II Soldiers Mutinied."

[16] Lindesay Parrott, "Pacific Veterans Press for Return," *The New York Times*, December 5, 1945, 6.

[17] Commander George W. Stedman Jr. to Coast Guard Commandant, *General A.W. Greely*, September 10, 1945, National Archives I, Washington, D.C.

[18] Stedman to Coast Guard Commandant, September 10.

[19] Jeffries, *Wartime America*, 170-171.

[20] "Baby Boomers," History, A&E Television Networks, June 1, 2021. https://www.history.com/topics/1960s/baby-boomers-1.

[21] "U.S.S. *General A.W. Greely* Deck Log," October 2 and October 3, 1945. National Archives 1, Washington, D.C.

Chapter 20

[1] I spoke with a representative of The Noble Maritime Collection in Staten Island, which has an exhibit of the *Andrea Doria* but no reference at that time to the *Tamaroa* in its list of participating rescue ships.

[2] "USCGC *Tamaroa* (WMEC-166)," Military History, n.d. https://military-history.fandom.com/wiki/USCGC_Tamaroa_(WMEC-166).

[3] "Missing Masterpiece," *Strange Inheritance*, FOX Broadcasting Company, February 6, 2017. https://www.fox.com/watch/af10908cc0a1866052bb07fb2e36f1e2.

[4] "Painting of Texas battle found in West Virginia attic," *Superior Telegram*, October 26, 2010. https://www.superiortelegram.com/lifestyle/arts-and-entertainment/painting-of-texas-battle-found-in-west-virginia-attic.

[5] Sam DeShong Ratcliffe (1992) *Painting Texas History to 1900*. Austin: University of Texas Press.

Bibliography

100 Years of Marriage and Divorce Statistics in the United States, 1867-1967. Human Resources Administration, National Center for Health Statistics, Rockville, Md., December 1973. https://www.cdc.gov/nchs/data/series/sr_21/sr21_024.pdf.

"1st Air Commando Group." Army Air Corps Library and Museum, n.d. https://www.armyaircorpsmuseum.org/1st_Air_Commando_Group.cfm.

"2nd Air Commando Group." Army Air Corps Library and Museum, n.d. https://www.armyaircorpsmuseum.org/2nd_Air_Commando_Group.cfm.

A Guidebook on Calcutta, Agra, Karachi and Bombay. The Publicity Division of the American Red Cross in the CBI theater, December 1943.

A Pocket Guide to Burma. War and Navy Departments, 1944.

A Pocket Guide to China. War and Navy Departments, n.d.

A Pocket Guide to India. War and Navy Departments, 1942.

Abel, Barbara. "Hail Hostess!" United Service Organizations Program Service Publication #522, n.d.

"Airliner Crashes in Bay near Boston." U.S. Coast Guard news release, October 24, 1960. https://catalog.archives.gov/id/205591464.

Albright, Alex. "African-American Navy Bands of World War II." Alexalbright.works, July 2022. https://alexalbright.works/research/music/african-american-navy-bands-of-world-war-ii

"An Iconic Chicago Landmark Comes to Life." Chicago Union Station, n.d. https://chicagounionstation.com/about/past.

Annual Report of the Postmaster General, for Fiscal Year Ended, June 30, 1942. Government Printing Office, Washington D.C., 1942.

"Baby Boomers." *History*, A&E Television Networks, June 1, 2021. https://www.history.com/topics/1960s/baby-boomers-1

Barnes, Tom. "Personality Studies Show the Difference Between People Who Play Music and Everyone Else." BDG Media, August 15, 2015. https://www.mic.com/articles/124032/personality-studies-show-the-difference-between-people-who-play-music-and-everyone-else.

"BB's Stage Door Canteen at The National WWII Museum." The National WWII Museum, New Orleans, n.d. https://www.nationalww2museum.org/events-programs/bbs-stage-door-canteen

"Beach Patrol XVII." The Coast Guard at War. Declassified document from Historical Section of the Public Relations Division of the U.S. Coast Guard, May 15, 1945.

"Becoming a Shellback *Greely* Style." Cartoon by Sgt. Locher published in *The Salt*, April 25, 1945.

"Bill Schallen." Discogs, n.d. https://www.discogs.com/artist/763816-Bill-Schallen .

"Biography of Assistant Director William F. Santelmann." The Official Website of Marines, n.d. https://www.marineband.marines.mil/About/Our-History/History-of-the-Assistant-Directors/William-F-Santelmann.

Bland, George. "Jammin' on Hatch 5." Cartoon published in *The Salt*, n.d.

Bown, Stephen R. "Scurvy: How a Surgeon, a Mariner, and a Gentleman Solved the Greatest Medical Mystery of the Age of Sail." Captain Cook Society, 2003. https://www.captaincooksociety.com.

"Capt. Stedman's Statement as the *Greely* Ends Maiden Trip." *The Salt*, May 20, 1945, Vol. 1, No. 32.

"Charles Darwin on Discovering Sea Sickness." *The Guardian*, September 26, 2016. https://www.theguardian.com/science/2016/sep/26/charles-darwin-discovering-seasickness-weatherwatch

Chicago Union Station. Wikipedia, n.d. https://en.wikipedia.org/wiki/Chicago_Union_Station

Chilton, Martin. "Shelly Manne, Remembering A Jazz Drumming Giant." *Udiscovermusic*, June 11, 2022. https://www.udiscovermusic.com/stories/shelly-manne-drummer-tribute

"Chinese Make Top Score in Duplicate Tournament." *The Salt*. Vol 2, No. 21, June 15, 1945.

"Clare E. Grundman." Altissimo Recordings, July 23, 2013. https://militarymusic.com/blogs/military-music/13516333-clare-e-grundman.

"Clare Grundman Archives." Ohio State University Libraries, n.d. https://guides.osu.edu/c.php?g=826137&p=5897960.

Cleveland, Les. *Dark Laughter: War in Song and Popular Culture*. Westport CT: Praeger Publishers, 1994.

"Cocos (Keeling) Islands." *The World Fact Book*, July 18, 2022. https://www.cia.gov/the-world-factbook/countries/cocos-keeling-islands.

"Commissioning Assault Cargo Ship U.S.S. *Theenim* (AKA-63)." Program for commissioning, December 23, 1944.

Connor, Joseph. "World War II Soldiers Loved to Sing—Provided They Got to Sing Their Way." Historynet, n.d. https://www.historynet.com/army-songs-in-world-war-2.

"Crowd-sourced: Norman Rockwell's Christmas at Chicago Union Station." *The Saturday Evening Post*, December 19, 2016. https://www.saturdayeveningpost.com/2016/12/crowd-sourced-norman-rockwells-christmas-chicago-union-station.

Daily Mirror. August 14, 1945.

Downey, Sally A. "Music on the Move." *Philadelphia Inquirer*. Nov. 5, 2007. https://mainlinetoday.com/life-style/music-on-the-move.

Dunlap, David W. "Photo That Was Hard to Get Published, but Even Harder to Get." *The New York Times*, March 28, 2013. https://archive.nytimes.com/lens.blogs.nytimes.com/2013/03/28/a-photo-that-was-hard-to-get-published-but-even-harder-to-get

"Editorial." *The Salt*. Vol. 1, No. 1, March 21, 1945.

"Eli Bublick Obituary." *Miami Herald*, February 5, 2012. https://www.legacy.com/us/obituaries/herald/name/eli-bublick-obituary?pid=155766602.

"Ernie Pyle." Indiana Historical Society, n.d. https://indianahistory.org/education/educator-resources/famous-hoosiers/ernie-pyle.

Ewing, W.M. "Sound transmission in sea water." Woods Hole Oceanographic Institution, 1941. https://darchive.mblwhoilibrary.org/server/api/core/bitstreams/2ed8bdd5-b577-59b4-9e42-76f7449ca363/content.

Ewing, W.M. and Worzel, J.L. "Long range sound transmission: interim report no. 1, March 1, 1944 - January 20, 1945," Woods Hole Oceanographic Institution, 1945. https://darchive.mblwhoilibrary.org/entities/publication/976cacef-a9ef-5db4-b3b4-e15232e4a267.

Executive Order 8985. https://www.presidency.ucsb.edu/documents/executive-order-8985-establishing-the-office-censorship.

"Flying Tigers Fought Japanese from China." National Museum of World War II Aviation, n.d. https://www.worldwariiaviation.org/flying-tigers-fought-japanese-from-china.

"Former Master Mariner Takes Over Coast Guard Cutter *Tamaroa*." *The New York Times*, February 13, 1952.

"From Luxury Liner to Troop Transport." *The Salt*, Vol. 1, No. 1, March 21, 1945.

"From Luxury Liner to Troop Transport: Story of a Skipper." *The Salt*, Vol. 1, No. 1, March 21, 1945.

"*General A. W. Greely* (AP-141)." Naval History and Heritage Command, July 10, 2015. https://www.history.navy.mil/content/history/nhhc/research/histories/ship-histories/danfs/g/general-a-w--ap-141.html

General Board meeting transcript. U.S. Navy, January 23, 1942. https://media.defense.gov/2021/Nov/30/2002900538/-1/-1/0/1942-Gb_Integration_Transcript.pdf

"Glorious, Illustrious Sons of Magellan." Certificate presented to Arthur Schnell for crossing the equator April 3, 1945. Dated June 20, 1945.

"*Greely* First Ship to have its own Band Aboard from the Start." *The Salt*, March 21, 1945.

"*Greely* Summer Cruise—April, May 1945." Cartoon published in *The Salt*, n.d.

"*Greely* War Diary," U.S. Navy. May 24, 1945. https://catalog.archives.gov/id/140057351.

"*Greely* War Diary," U.S. Navy. July 1945. https://catalog.archives.gov/id/83528214.

"*Greely* War Diary." National Archives, August 1945. https://catalog.archives.gov/id/83528214

Guise, Kimberly. "Mail Call: Letters from the Archives." The National WWII Museum, March 7, 2018. https://www.nationalww2museum.org/war/articles/mail-call-letters-archives.

Guise, Kimberly. "Mail Call: V-mail." The National WWII Museum, New Orleans, December 7, 2019.

Harri, Katlin. "The Three Major Shifts in Soviet Music During World War II." Doclib, n.d. https://docslib.org/doc/9245658/the-three-major-shifts-in-soviet-music-during-world-war-ii.

Hinnant, Commander J.R. to Schoen, Mrs. Frank E. Washington, D.C. July 15, 1945. National Archives II, College Park, Md.

Hinton, James. "The Sinking of the S.S. *Normandie* at NYC's Pier 88." *New York Almanack*, September 23, 2014. https://www.newyorkalmanack.com/2014/09/the-sinking-of-the-s-s-normandie-at-nycs-pier-88.

Hogan, David W. "U.S. Army Special Operations." Center of Military History Department of the Army, 1992.

"History of the United States Coast Guard Band." U.S. Coast Guard documentary, n.d. https://www.youtube.com/watch?v=BZpwp0Sz3zc.

"Hollywood and the Office of War Information, 1942-1945." Backlots, February 19, 2020. https://backlots.net/2020/02/19/hollywood-and-the-office-of-war-information-1942-1945.

"How a life jacket came to be named after Mae West." *Word Histories*, 2018. https://wordhistories.net/2018/03/09/mae-west-life-jacket.

"How Five Brothers Perished at Sea and Inspired an American Legend." The Sullivan Brothers, July 6, 2020. https://thesullivanbrothers.com/how-five-brothers-perished-at-sea-and-inspired-an-american-legend.

"Initiate Greelytes into 'Sons of Magellan' Round the World Order." *The Salt*. Vol. 8, No. 27, July 21, 1945.

"Instrument of Surrender." Records of the U.S. Joint Chiefs of Staff; Record Group 218, September 2, 1945. National Archives. https://catalog.archives.gov/id/1752336.

"Jail for the Faithless." *Newsweek*, August 13, 1945. Found in http://www.oldmagazinearticles.com/ww2-divorces-pdf 26.

Japanese Language Guide. War Department, 1943.

Jeffries, John W. *Wartime America: The World War II Home Front*, Ivan R. Dee Inc., 1996.

Kellar, McKenzie. "Songs for America: Patriotism through Music." *Music Politics*, University of Alabama College of Arts & Sciences, 2017. https://www.musicpolitics.as.ua.edu/projects/primary-source-projects/songs-for-america-patriotism-through-music

Klugewicz, Stephen M. "Ten Great American Civil War Songs." *The Imaginative Conservative*. September 1, 2016. https://theimaginativeconservative.org/2016/09/ten-great-american-civil-war-songs-stephen-klugewicz.htm.

Koop, Theodore F. *Weapon of Silence*. University of Chicago Press, 1946.

Lange, Greg. "President Truman announces V-E Day, victory-in-Europe, on May 8, 1945." History Link, 1999. https://www.historylink.org/File/1279.

"Le Havre: World Heritage Site: The Bombing of 1944." UNESCO, n.d. http://unesco.lehavre.fr/en/understand/the-bombings-of-1944

Lee, R. Alton "The Army Mutiny of 1946." *The Journal of American History*, Oxford Academic, Vol. 53, Issue: 3, November 30, 1966, 555-571.

Litoff Barrett, Judy and Smith, David C. "'Will he get my letter?' Popular Portrayals of Mail and Morale During World War II." *Journal of Popular Culture*, Spring 1990, 32:4.

"Loose Lips do Sink Ships." World War II in the Pacific, October 12, 2001. http://www.ww2pacific.com/congmay.html.

Lotchin, Roger. "Mobilization for the Duration: The Bay Area in the Good War." National Park Service, n.d. https://www.nps.gov/articles/000/mobilization-for-the-duration-the-bay-area-in-the-good-war.htm?utm_source=article&utm_medium=website&utm_campaign=experience_more&utm_content=small.

New World Telegram. June 7, 2045. 1.

New York Journal American. August 21, 1945.

New York Journal American. August 6, 1945. 1.

New York Journal American. August 8, 1945. 1.

New York Post. August 14, 1945.

New York World Telegram. April 13, 1945. 1

Newcomb, Richard. *Abandon Ship!: The Saga of the U.S.S.* Indianapolis, *the Navy's Greatest Sea Disaster.* Henry Holt and Company, 1958.

Nieves, Edwin and Kinane, Sean. "The Long Blue Line: WWII recruit training at Manhattan Beach and desegregating the U.S. military," United States Coast Guard, February 10, 2023. https://www.mycg.uscg.mil/News/Article/3259454/the-long-blue-line-wwii-recruit-training-at-manhattan-beach-and-desegregating-t/.

McIndoe, Dominique. "Married men to get draft deferment—for now." World War 2.0, June 26, 1942. https://blogs.shu.edu/ww2-0/1942/06/26/married-men-to-get-draft-deferment-for-now

McLuhan, Marshall. *Understanding Media: The Extensions of Man.* Cambridge: The MIT Press, 1964.

McPartland, John. "Our changing Sexual Code: Endless Variety in Stories and Pictures." *Coronet* magazine, August 1955. http://www.oldmagazinearticles.com/article-summary/sex_during_ww2#.YtbHfYTMKUm.

Miller, Christine. "Playing a Role in the War: Music in World War II." Concordia Memory Project of Concordia College, n.d. https://concordiamemoryproject.concordiacollegearchives.org/exhibits/show/sartyessays/christinemiller.

"Missing Masterpiece." Strange Inheritance. FOX Broadcasting Company, February 6, 2017. https://www.fox.com/watch/af10908cc0a1866052bb07fb2e36f1e2.

More Fun en Route for our Armed Forces, National Recreation Association Inc., 1944.

Mowis, I.S. "Dick Stabile Biography." IMDb Mini Biography, n.d. https://www.imdb.com/name/nm0820895/bio?ref_=nm_ov_bio_sm.

"Music is a Strong Builder of Morale, Station's Touring Unit Discovers." *Harpoon,* Vol. II, No. 9, July 1, 1943.

Musser, James. "Tenth Air Force (AFRC)." January 13, 2019. https://www.afhra.af.mil/About-Us/Fact-Sheets/Display/Article/433512/tenth-air-force-afrc.

"Naval Station, Treasure Island." California State Military History and Museums Program, n.d. https://www.militarymuseum.org/NSTI.html

"New System for Overseas Mail." *Sparkles* newsletter, Vol. 1 No. 29, November 1, 1944. https://media.defense.gov/2022/Feb/25/2002945223/-1/-1/0/1944_Nov_1_Spars_Sparkles_ewsletter.pdf.

"Ninth Air Service Chief Demoted, Sent Home, for Talk About D-Day; Maj. Gen. H.J.F. Miller Named in Washington as 'Broken' by Eisenhower for Remark at London Cocktail Party." *The New York Times,* June 8, 1944, 4. https://www.nytimes.com/1944/06/08/archives/ninth-air-service-chief-demoted-sent-home-for-talk-about-dday-maj.html.

"Non-Judicial Punishment/Article 15/Captain's Mast." Karns Law Firm, n.d. https://www.usmilitarylawyer.com/military-non-judicial-punishment-article-15.asp

"Offer of Loan of Musical Instrument," Undated, unsigned Coast Guard contract. National Archives II, College Park, Md.

"Our Exec. Has Record of 33 Years at Sea." *The Salt,* Vol. 1, No. 1, March 21, 1945/.

"Out of Wedlock Births." *Pathfinder,* Dec. 11, 1944. http://www.oldmagazinearticles.com/

WW2-illiegitimate-births-pdf.

"Painting of Texas battle found in West Virginia attic." *Superior Telegram,* October 26, 2010. https://www.superiortelegram.com/lifestyle/arts-and-entertainment/painting-of-texas-battle-found-in-west-virginia-attic.

Parrott, Lindesay. "Pacific Veterans Press for Return." *The New York Times,* December 5, 1945.

Perkins, Sam. "Why World War II Soldiers Mutinied after V-J Day." *History,* 2022. https://www.history.com/news/world-war-ii-soldiers-mutiny-v-j-day

"Plane crashes into Empire State Building." *History,* A&E Television Networks, November 13, 2009. https://www.history.com/this-day-in-history/plane-crashes-into-empire-state-building/.

Potter, Pamela M., Baade, Christina, and Marvin Montemorra, Roberta. *Music in World War II,* Bloomington: Indiana University Press, 2020.

"Pullman Porters." History, A&E Television Networks, October 8, 2021. https://www.history.com/topics/black-history/pullman-porters.

Quesada, Alejandro de. *U.S. Coast Guard in World War II.* New York: Osprey Publishing, 2011. Kindle.

Ratcliffe, Sam DeShong. *Painting Texas History to 1900.* Austin: University of Texas Press, 1992.

Reed-Hill, Ellis. Letter from the Coast Guard Public Relations Office to Edward Lloyd, editor, *U.S. Coast Guard Magazine.* January 8, 1942. Archives I, Washington, D.C.

"Report of the Sinking of the SS President Coolidge." *Theleansubmariner,* December 12, 1942. https://theleansubmariner.com/2012/12/12/december-12-1942-report-of-the-sinking-of-the-ss-president-coolidge.

"Revealed: What really happened when Glenn Miller disappeared in 1944." University of Colorado Boulder, College of Music, July 8, 2014. https://www.colorado.edu/music/2014/07/08/revealed-what-really-happened-when-glenn-miller-disappeared-1944.

Richards, Phil, and Banigan, John. *How to Abandon Ship.* Fighting Forces Series. Cornell Maritime Press, 1943.

Romm, Cari. "During World War II, Sex Was a National-Security Threat." *The Atlantic,* October 8, 2015. https://www.theatlantic.com/health/archive/2015/10/during-world-war-ii-sexually-active-women-were-a-national-security-threat/409555.

Schoen, Mrs. Frank E., Brooklyn, N.Y., to King, Ernest J., admiral and commander of the U.S. Fleet. June 27, 1942. National Archives II, College Park, Md.

Shaw, Christopher C. "On the Dynamics of motion sickness on a seaway." *The Scientific Monthly,* Vol. 78, No. 2, February 1954, 110-116. https://www.jstor.org/stable/21192.

"Shellback or Pollywog: The US Navy's Line Crossing Ceremony Revealed." *Navy Crow,* July 7, 2022. https://navycrow.com/shellback-the-us-navys-line-crossing-ceremony-revealed.

"Shelly Manne music." Cartoon published in *The Salt,* n.d.

"Sixth War Loan." *The New York Times,* December 19, 1944.

"Soldiers Relate Tales of Naga-Head-Hunters." *The Salt,* Vol. 2, No. 21, June 15, 1945.

"Statement by the President on the Death of Ernie Pyle." Harry S. Truman Library, Museum (National Archives, April 18, 1945). https://www.trumanlibrary.gov/library/public-papers/6/statement-president-death-ernie-pyle.

Stedman, George W. Jr., Memo September 19, 1945. National Archives II, College Park, Md.

Stedman, George W. Jr. to Coast Guard Commandant, U.S.S. *General A.W. Greely* Personnel; report on September 10, 1945. National Archives I, Washington, D.C.

Stewart, G.U., Manhattan Beach Commanding Officer to Commandant, Manhattan Beach, N.Y., September 29, 1945. National Archives II, College Park, Md.

"Suits for Divorce," *The Albuquerque Tribune,* August 23, 1961. p. 21.

Suits-Smith, Kathleen. "He feels competition is large part of life," *Evening Telegram,* October 22, 1966.

"Tars and Spars, the Coast Guard Show." Playbill, Artcraft Litho and Printing company, New York, n.d. https://www.history.uscg.mil/Portals/1/images/faq/tars_and_spars. pdf?ver=2017-07-11-153815-170.

"Tenth Air Force Overview: 10 March 1942 - December 1945, US Army Air Forces in China Burma India." n.d. https://www.aafincbi.com/CBI_10af_home.html.

Tewksbury, R.S., "Plan of the Day." Memo to *Greely* crew, April 7, 1945.

Thiesen, William H. "The Coast Guard's World War II Crucible." Naval History Magazine, Volume 30, Number 5, October 2016. https://www.usni.org/magazines/naval-history-magazine/2016/october/coast-guards-world-war-ii-crucible.

Thiesen, William H., "First to serve, first to fight and first to sacrifice—African Americans in the U.S. Coast Guard." MyCG (Feb. 2, 2024). https://www.mycg.uscg.mil/News/Article/3664598/.

Trussell, C.P. "Deferring of Married Men in Draft Is Written into Allowances Bill." *The New York Times*, June 13, 1942, 1.

"The Coast Guard Sings." Playbill of the United States Coast Guard, Third Naval District. New York, November-December 1944.

"The history of the battle song Lili Marlene sung by Lale Andersen." Panamint Cinema, n.d. https://www.panamint.co.uk/the-true-story-of-lili-marlene-dvd/.

"The Ledo Road." CBI-Theater (2022). http://www.cbi-theater.com/ledoroad/Ledo_Main.html/.

"The Legacy of Ithaca College Bands." Ithaca College, n.d. https://www.ithaca.edu/academics/school-music/ensembles/band/history-bands-ic.

"The Sullivan Brothers: Transcripts of Service." Naval History and Heritage Command, November 7, 2017.

The Valhalla Junior High. *Yearbook of Valhalla Junior High School*, 1931.

"Troops and Crew Rate the *Greely* Grenadiers '4.0.'" *The Salt*, August 23, 1945.

U.S. Census, 1950.

U.S. Coast Guard Official Dispatches. January 3, 1945. January 10, 1945, January 23, 1945, March 3, 1945.

"U.S. Office of War Information." Franklin D. Roosevelt Presidential Library and Museum, n.d. https://fdr.artifacts.archives.gov/people/6707/us-office-of-war-information/objects/.

"U.S.S. *Etamin*, AK-93." United States Coast Guard, January 4, 2021. https://www.history.uscg.mil/Browse-by-Topic/Assets/Water/All/Other-Vessels-Non-CG/Article/2461253/uss-etamin-ak-93.

"U.S.S. *General A.W. Greely* Deck Logs." March 1945-December. National Archives I. Washington, D.C.

"U.S.S. *Greely*." Undated newsletter published on the U.S.S. *Greely* during the first voyage, n.d.

"USCGC Tamaroa (WMEC-166)." Military History, n.d. https://military-history.fandom.com/wiki/USCGC_Tamaroa_(WMEC-166)

Vento, Carol Schultz. "Censorship and World War II." Defense Media Network, July 13, 2014. https://www.defensemedianetwork.com/stories/censorship-and-world-war-ii

"Vt. National Guard band gets war recognition." Boston.com. January 16, 2012. http://archive.boston.com/news/local/vermont/articles/2012/01/16/vt_national_guard_band_gets_war_recognition/#:~:text=The%2040th%20Army%20band%20shipped%20out%20to%20the,to%20U.S.%20forces%20on%20the%20Island%20of%20Guadalcanal.

Young, William H., and Young, Nancy K. *Music of the World War II Era*. Westport, CT: Greenwood Press, 2008.

Walton, William. "The Spencer gets her Sub." *Harpoon*. U.S. Coast Guard Training Station, Manhattan Beach. Vol. II, No. 9, July 1, 1943.

"War correspondent Ernie Pyle killed." *History*, April 15, 2020. https://www.history.com/this-day-in-history/journalist-ernie-pyle-killed.

"War Damage to the U.S.S. *Etamin*, Preliminary Report of." U.S. Navy, May 1, 1944. https://catalog.archives.gov/id/134367013?objectPage=2.

"War Damage U.S.S. *Etamin*, Final Report of." U.S. Navy, June 1, 1944. https://catalog.archives.gov/id/78477663?objectPage=3.

"What causes seasickness?" National Ocean Service, National Atmospheric and Oceanic Administration, n.d. https://oceanservice.noaa.gov/facts/seasickness.html. Accessed September 25, 2022.

"World War II in San Francisco, California." *Legends of America*, n.d. https://www.legendsofamerica.com/san-francisco-world-war-ii.

"World War II Shipbuilding in the San Francisco Bay Area." National Park Service, n.d. https://www.nps.gov/articles/000/world-war-ii-shipbuilding-in-the-san-francisco-bay-area.htm.

Zobenica, Jon. "The Greatest Sexual Revolution--How World War II Prefigured the '60s." *The American Scholar*, December 2, 2019. https://theamericanscholar.org/the-greatest-seual-revolution.

Appendix

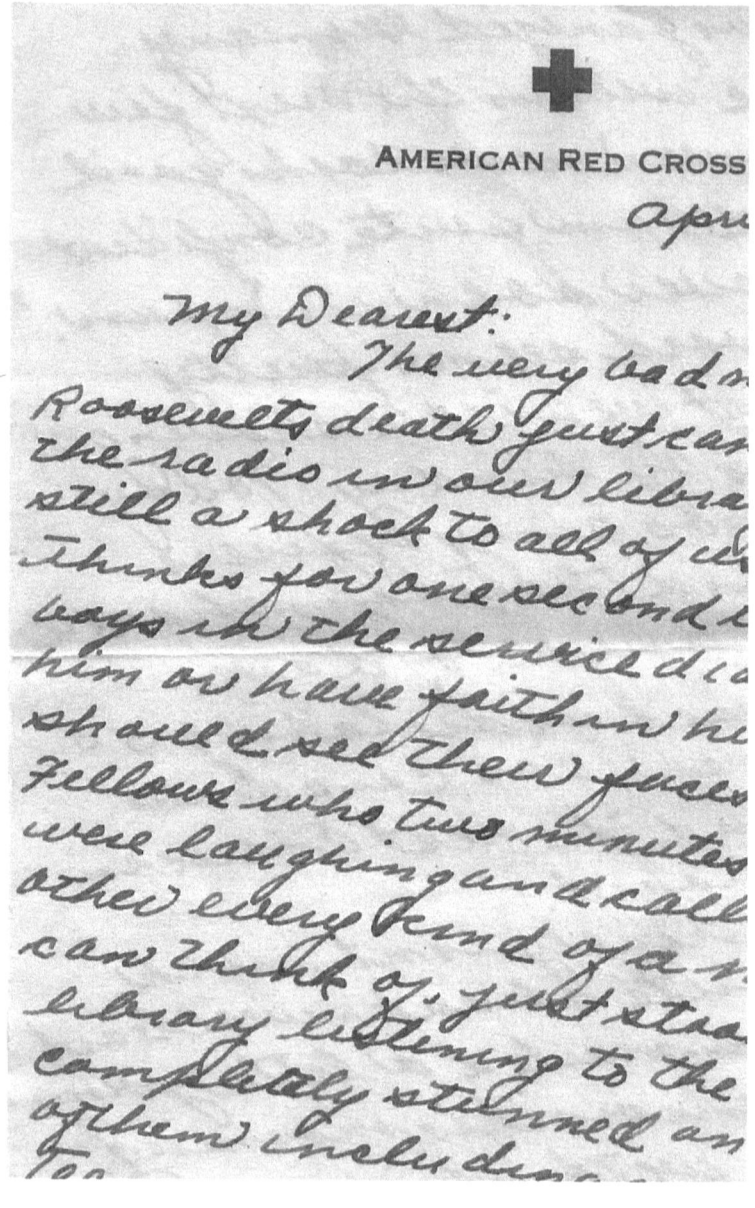

213 • Larry Schnell

Fr

at

My Darling-

Happy day! I n

thought I'd see it

is wonderful. We

days holiday - au

rationing is over.

peoples bought g

out yesterday that

wasn't

park. near Jama
Honey you would
love it - its so
and beautiful. t
- All we wished
yesterday were du
Everything else
wonderful. The
was nice & the
so blue - I eve
nature was ce
yesterday. Whee

and of course I'm
to day. Hey Sunny
have to be nice
The war's over a
besides I've got a
coming. I can't n
see you — you a
Mousie. Please hur
we can celebrate
aint life —— wonderf
— "We have as yet receive
your cute letters from Joe
only bright spots.

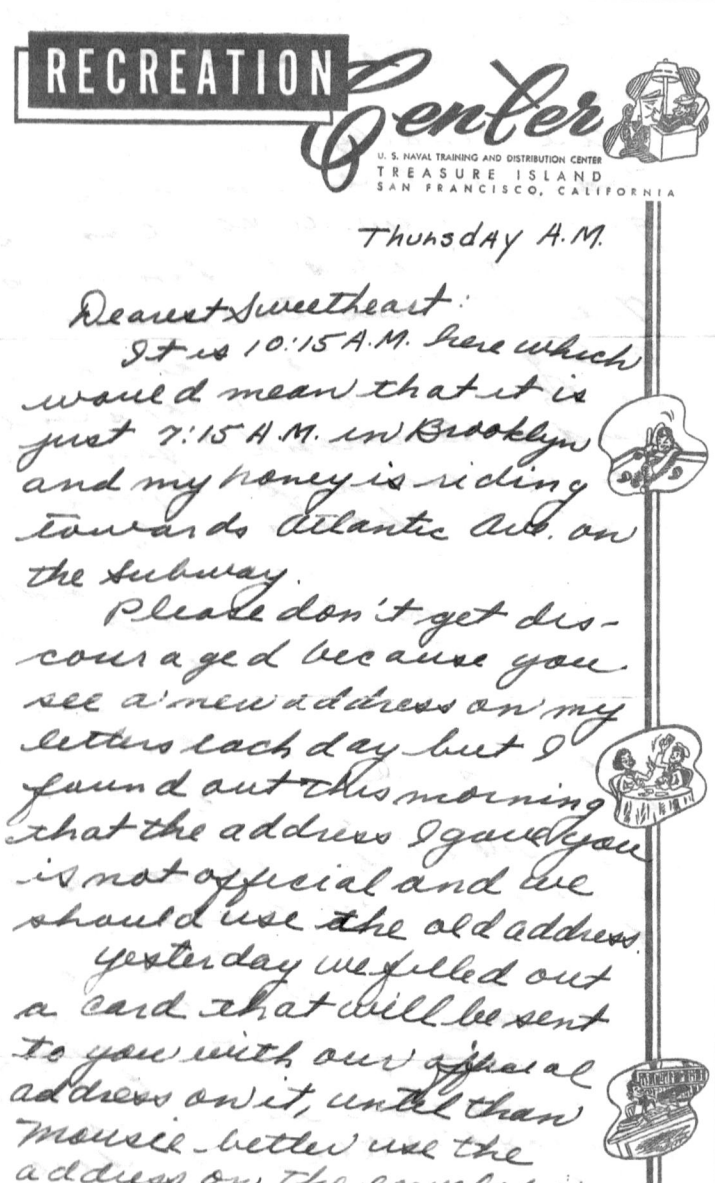

RECREATION *Center*

U. S. NAVAL TRAINING AND DISTRIBUTION CENTER
TREASURE ISLAND
SAN FRANCISCO, CALIFORNIA

Thursday A.M.

Dearest Sweetheart:

It is 10:15 A.M. here which would mean that it is just 7:15 A.M. in Brooklyn and my honey is riding towards Atlantic Ave. on the subway.

Please don't get discouraged because you see a new address on my letters each day but I found out this morning that the address I gave you is not official and we should use the old address.

Yesterday we filled out a card that will be sent to you with our official address on it, until then mousie – better use the address on the envelope.

The other address may reach us but just to be sure use this one.

We finally met our Che today and he seems like a good fellow. Also the ship chaplain was talking to us. The ship has two pianos a big recreation hall, a li with over 2000 books in and a Hammond Organ. It really sounds pretty nice me. The chaplain also tol us that the crew is mad up entirely East Coast an southern Boys and that should Germany collap within the next couple o months we may run between Europe and N.Y

I have, with Boebertz hel and mostly Franks wa figured out a code to let you know whe...

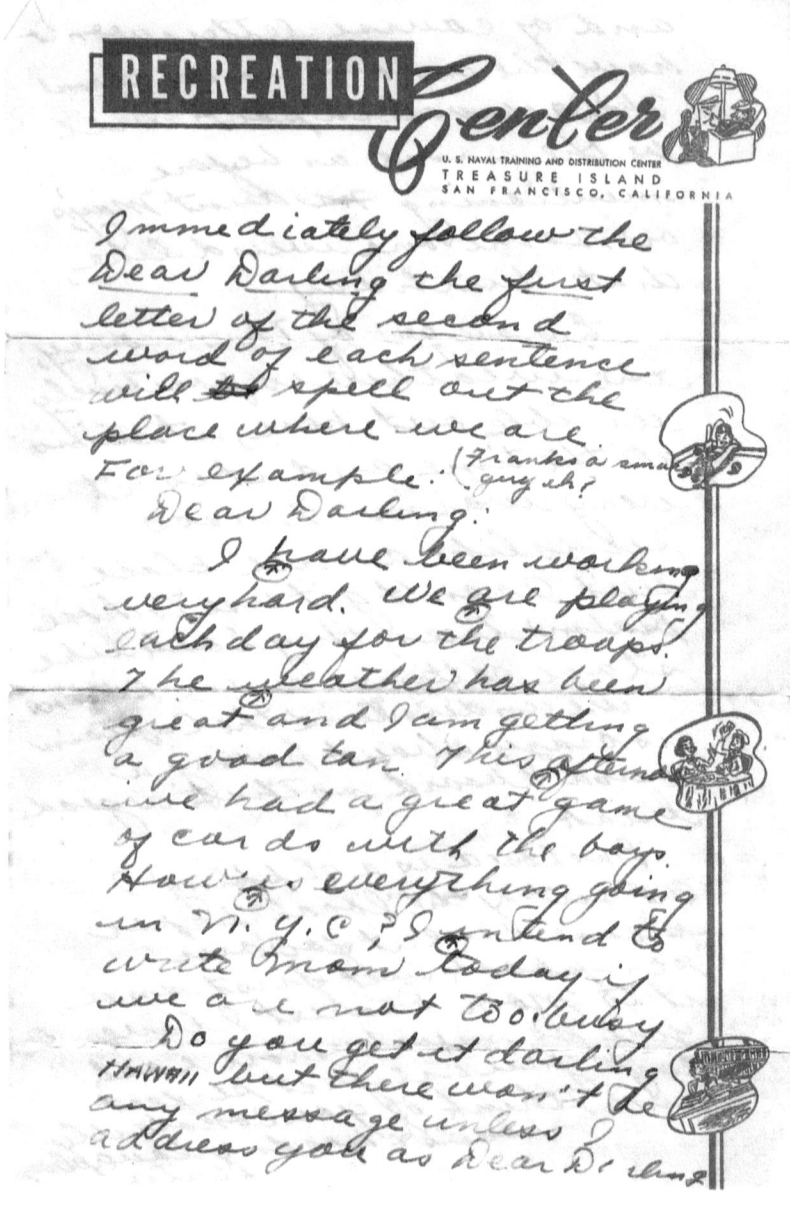

RECREATION Center

U. S. NAVAL TRAINING AND DISTRIBUTION CENTER
TREASURE ISLAND
SAN FRANCISCO, CALIFORNIA

Immediately follow the
Dear Darling the first
letter of the second
word of each sentence
will spell out the
place where we are.
For example. (Franks a smart
guy eh?
Dear Darling.
    I have been working
very hard. We are playing
each day for the troops.
The weather has been
great and I am getting
a good tan. This afternoon
we had a great game
of cards with the boys.
How is everything going
in N.Y.C.? I intend to
write mom today if
we are not too busy
    Do you get it darling
HAWAII but there wasn't de
any message unless I
address you as dear Darling

A.R. Schnell M[...]
U.S.S. New. a.W.[...]

Air Mail

Wednes[...]
May 2,

Dear Darling —

. They made an ann[...]
tonite over the loud spea[...]
that Hitler had just been [...]
streets of Berlin. Guess eve[...]
mighty happy about that [...]
to be in N.Y.C. now to se[...]
celebration. Our being u[...]
here, the European war [...]
very remote, at least to [...]

If only you could [...]
Bland and I wish mu[...]
money. As usual I expec[...]
objection from you dea[...]
rather like it now. Its [...]
cool ~~~~ and we had [...]

A. R. Schnell MM
U.S.S. Gen. A. W. S

②

Don't forget to let me kn
ow you any Weny are gett
along, I hoped. Its really
um being away from y
arling, each day I get
onesome. If I am not m
here are a lot of loneso
mys out here. I doubt y
eally appreciated how
ell. It was to have our ou
here are so many things I
oue to tell you but can
ecomes gecete disgusting.
ould only see Eli Bublu
board ship you wou
aughing at some of th

A. R. Schnell Mus. 2/c
U.S.S. Gen. A. W. Gree

③

ne more letter before u
t a' port and that w
l you will get for aus
ter tomorrow no mail u
out till we get to our
ort so don't worry dea
u should get about fif
tters in this group.

I am feeling fine seve
nd every thing is going
ong nicely, say hel to t
mers and Fran for m
u are so precious to m
a, I think about you
ring the day and dr
you at nite. nobody
ed ever to n...

About the Author

Larry Schnell is a writer and journalist living near Cooperstown, New York. As a reporter for the *Gainesville Sun* and the *Florida Times-Union*, Larry Schnell covered business, politics, criminal justice, environment, and higher education. He graduated from the State University of New York at Oneonta with bachelor's degree and from the University of Florida with a master's degree. He has taught English, journalism, and writing for several colleges and the University of Florida and published three books of fiction, *Country People*, *The Year of the Gator*, and *The Dry Tortugas*.

A lifelong sailor and licensed sailboat captain, Larry Schnell has sailed in Florida, the East Coast, the Bahamas, and Lake Ontario in his sailboat *Intrada*, and has published articles in boating magazines about his sailing experiences.

Larry Schnell can be reached at larryschnellfl@yahoo.com

Visit Larryschnell.com for additional information.

Dear Reader,

If you enjoyed this book, please consider posting a short review on Amazon, or wherever you purchased this book. Reviews help authors a great deal, and help us continue creating new books for your enjoyment. Thank you!

ALSO FROM INTRADA PRESS LLC

Country People

The Year of the Gator

The Dry Tortugas

Available through amazon.com, Barnes and Noble, and by request through most major and independent book sellers.
Author Larry Schnell is available for presentations and discussions about A Sailor's Song: Lost Love Letters of World War II.

www.ingramcontent.com/pod-product-compliance
Lightning Source LLC
Chambersburg PA
CBHW031504120626

46545CB00005B/1737